Practical Manual of
Fisheries

The Author

Dr. K.P. Biswas, M.Sc., Ph.D., D.F.Sc. (Bombay), E.F. (West Germany), F.Z.S., F.A.B.S. (Kolkata), Former Joint Director Fisheries (L-I), Government of Odisha, Director Fisheries, Andaman and Nicobar Islands, Government of India, Principal Fisheries Training Institute, Fishery Technologist (ICAR) and at present Faculty Member of Marine Science Department, University of Calcutta and Faculty Member, Department of Fishery Engineering, West Bengal University of Animal and Fishery Sciences has published twenty four books and more than hundred research and review papers on Fisheries, Marine Sciences and Aquatic ecology. His last book, entitled, "Breeding and Hybridization of Food Fishes" has been published in 2016.

Dr. Biswas received Life Time Contribution Award, 2016 in appreciation of his long dedicated and outstanding contributions in the field of Zoology from The Zoological Society, Kolkata on July 29, 2016.

Practical Manual of
Fisheries

– Author –

Dr. K.P. Biswas

2018

Daya Publishing House®

A Division of

Astral International Pvt. Ltd.

New Delhi – 110 002

ISBN 9789387057746 (International Edition)

Publisher's Note:

Published by : **Daya Publishing House®**
A Division of
Astral International Pvt. Ltd.
– ISO 9001:2015 Certified Company –
4736/23, Ansari Road, Darya Ganj
New Delhi-110 002
Ph. 011-43549197, 23278134
E-mail: info@astralint.com
Website: www.astralint.com

Dedication

**Dedicated to Mrs. Maju Biswas
for her interest in the book**

Acknowledgement

The author deeply acknowledge the help of Dr. N.A. Talwar for preparing the print out of hard copy of the manuscript.

K.P. Biswas

Preface

Fisheries work, especially in practical field embraces many variations. Important among them are fish and aquatic production, both in natural waters, like rivers, streams, lagoons, sea and estuaries and also in man-made impoundments, like, ponds, tanks, irrigation reservoirs, pens and cages. Their ecological environments are different from moving waters to static condition. Their ecosystems need to be explored and monitored on a regular basis to obtain sustainable and increased production.

Beginning with the inventory of the ecosystem and monitoring them regularly generate data, that can be used for planning and development with an object to raise the productivity.

Perturbing the ecosystem with pollutants need special steps for inventory and thereafter monitoring on a regular basis to assess the condition of the ecosystem after enforcing measures for restoration.

Currently hundreds of methods exist for the collection, analysis and interpretation of aquatic biological data. There are some of the standard references. In addition, there are numerous other "methods" publications that are appropriate for selected groups of organisms or for use in particular habitat types. This is especially true of aquatic life studies, in higher gradient and upland streams where impact of torrential waters and mineral mining is severe.

Aquatic communities, not like terrestrial ones often contain numerous species whose abundance range from common to very rare. Those various species abundances reflect complex shifts in biotic and abiotic factors, which ultimately affect the fitness of each community member. In addition, immigration and emigration may continuously affect species composition, further complicating the biological inter-actions taking place. It is in this regard therefore, that a structural approach of studing aquatic communities, namely, examining the type, number, and abundance of the species present must be viewed as a dual aspect of the dynamics of the behavior of those communities. A great deal of ecological theory has arisen around this concept and numerous techniques have been proposed for obtaining descriptive measures of community structure as indication of more complex functional characteristics. Debate over interpretation of these measures, however, is far ranging among ecologists and will doubtless continue until a more thorough understanding of community function is obtained.

Despite this difficulty, descriptive measurements of community structure, particularly with respect to species number and abundance, have proved to be more powerful tools with ichthyologist to summarize large amount of data its capability in turn facilitates comparison between different communities with respect to both time and space. As studies of community functional process emerge, this economy of description has obvious advantages for establishment of any regularities existing between structure and function, thereby providing more realistic and efficient means of managing aquatic resources.

In this book an attempt has been made to formulate an uniform guide line for the different methods of practical field work in fisheries.

K.P. Biswas

Contents

Introduction

Methods of obtaining observations on components of the ecosystem are frequently called sampling methods. Usually such a " sample" is a "grab sample" in the terminology of sampling science, and it is often useful to treat this as an observation in a more generally structured sampling scheme.

But methods of obtaining observational "samples", the possibilities and their inherent properties are often critical. Traps, electro-shockers, kick samples, call counts, tracking stations, flushing counts and a great many others are valuable techniques for sampling the biota.

In scientific sample, a population which is distributed in time and space will require some control over the "sampling", if one wishes to "represent" the population by a sample. Characteristically the scientific sample is identified as "a set of observations distributed in time and space in such a manner that the population (or the relevent population behavior) is represented". Each observation in this sample may be a "sample" in the sense of representativeness with the least restriction which must be used for inference about the population.

Probability sample is one conducted in such a manner that the formal theory of scientific sampling can be applied. The statistical properties of the estimation are known, for example, it is usual that a probability sampling leads to consistent estimators. Sometimes unbiased estimators exist and are useful. But very often it is the case that one needs only a sampling scheme, which provides subjective reoresentativeness, in such a case, a probability sample is unnecessary.

A scientific sampling program is properly designed to answer scientific questions. It is always advantageous to express these questions explicitly prior to conducting the sampling.

Some kinds of relevent questions are not addressable by a formal scientific sampling program, but require instead a scientific study, engaged by an investigator, or a team of investigators. Such study may generate sub-questions, which are answerable by a sampling program.

It sometimes happen that a data set which has generated for one purpose, turns out to be useful for another, but there is no way to gurantee this in advance. That is, an anticipated question can not be guranteed to be answerable from the data currently being collected. This has several important implication.

So long as new questions are being asked, the base line data which exist at any point of time will be seem to be inadequate.

One must attempt to anticipate questions in order to establish the data base for their answer.

Implication of these points for a program of impact assessment and monitoring are several and significant. The program must be viewed as fulfilling several quite different roles at several levels of scientific study and several horizons of planning and data accumulation.

A general scientific investigation of the impacts of the pollution sources of concern on the ecosystems of concern must underlie the entire effort. It is not carried on within the program, it must be closely connected by an external structure. This investigation must establish ongoing studies of impacts to stay abreast of newly arisen concerns and maintain sound basic research in the structure and function of the involved ecosystems. The study of perturbed ecosystems, in the context of natural ecosystems, demand the most modern and sophisticated research program. But this is not a service program, it is basic research at its best and needed for the vitality of the program dealing with impacts.

The second facet of the impact investigation program must be an inventory and monitoring program. This is generally involved with standard inventory, but also with establishment of baseline data sets and system description as well as monitoring for surveillance, compliance and for assessment of foregoing impacts. This is primarily a service function, but the identification of monitoring variates must be current and tied closely to the scientific study in order to take advantage of new structure and theory.

Analysis and assessment of impact of pollution sources makes use of the theory developed by the scientific program and the data provided by the inventory and monitoring program, to appraise the impact. The purpose of the separation is to maintain the distinction between ecological impacts and social impacts, or the value to the society of the ecological impacts. The models and studies of the scientific program used with the data of inventory and monitoring program generate the assessment of biological impacts. These are then filtered appropriate socio-economic model to generate the value of those impacts.

Statisticians complain that far too often masses of scientific data are accumulated, as a routine basis without any clear objective in view. Research workers in fish culture frequently believe that, because certain routine observations are commonly made in general limnological studies, they must necessarily be made in experiments on fish culture. Quite often no new knowledge is obtained from those data and time and energy could be better utilized.

Many research papers give the impression that the authors have launched experiments without any clear understanding of the problems they are trying to solve. Formulating the problem properly is the first step, and time spent on this is time well spent.

K.P. Biswas

Chapter 1

Standard Statistical Methods for Fish Culture Experiments

Formulating the Problem

It is essential that statistical methods should be applied right from the beginning of an experiment, and not as is so often the case at the end. The experiment must be designed to obtain data which can be analyzed statistically. This is best done with the help of a statistician, but one is not always available in a fish culture research center, the research worker must acquire adequate proficiency in the design of experiments.

To better utilize time and energy in carrying out observations it is important to decide the minimum number of variables which must be studied to give adequate data for the solution of the problem under investigation. Equally important is the number of observations which must be made on each variable to give data which have statistical meanings. For instance, comparison between two ponds receiving different treatments may give a useful pointer, but the data may be too limited for statistical analysis.

If ponds are to be constructed specially for the experiments they can be designed to permit several variation to be studied at the same time. If however existing ponds are to be used, it may be possible to vary only one factor at a time, and variables will therefore have to be studied sequentially; it is important in this case to decide the order of priorities. Unlike agricultural experiments, where land can be ploughed and different arrangements of plots designed, very little changes can be effected in ponds and this factor governs the experimental design, since any treatment may give a permanent bias in the properties of the ponds. For example, in places it might have been better if, instead of introducing phosphate, nitrogen and potash, in the first trial, only phosphate had been tested at several levels. This would have demonstrated the over-fertilization effect. Nitrogen and potash could have been introduced in the second or third trials, and it would have been easily found that they were not really necessary.

Design of Experiment

The age effect – As soon as a pond is filled with water it undergoes chemical, physical and biological changes, and these are irreversible. The previous history of a pond is therefore of great importance. Ideally, when several ponds have to be used, they should be constructed at the same time and thus be of the same age. Even if this is done it may not be possible to fill them all at the same time and a program must be worked out to reduce the variance due to the time of filling. In Malacca, in the Latin square consisting of 16 ponds of 0.405 ha each, three ponds are filled each week and they are stocked with fish, the fertilizers applied and finally harvested in the same order. Thus all ponds have water, fish and fertilizer for the same length of time. Variations can occur with the time of filling, as can be seen by the gradual shift of plankton succession in Malacca, but this shift has had little effect on the final crop of fish. In Malacca river water was used, but with artesian water even this variable could be eliminated. If ponds cannot be constructed at the same time, experiments should be designed to reduce the variance due to the age of the ponds. If the differences in the latter are wide, the design must permit accurate statistical assessment of variance due to this.

Effect of Experimental Treatment

It is an easy and common approach in fish culture research to await the results of one trial before deciding on the treatment for the next. This method of approach has hidden dangers. Fish ponds are usually shallow, and so the bottom mud may exert a marked effect on the fertility and biota of the overlying water. Any treatment of the pond, particularly the application of fertilizers, may produce a permanent bias in the properties of the bottom mud. If a series of experiments is to be carried out, the sequence should be planned as a whole.

Testing a Hypothesis

There may be no harm in having an idea as to what the outcome of an experiment will be, but the experiment must be designed to test a hypothesis and not to prove it. It may be possible to support a theory by arranging the experiment with a bias, even if this is done unconsciously. For example, ponds with least water loss due to seepage may be selected for the application of fertilizers, while it might not be considered necessary for controls. Statistical design eliminates this bias, hence the use of randomized plots or Latin squares.

Standardization of the Fish

Unless the optimum stocking rate is the object of experiment, fish should be stocked at the same rate. In fertilizer trials the ponds should be of the same size and have the same number of fish in each. In a fertilizer trial in the Philippines (Padlan, 1956) the validity of the conclusion that the fertilizer used was beneficial was made doubtful because 2500 fish were stocked in the pond with fertilizer as against 1000 in the control. If the crop of fish in the fertilized pond (9484 kg) is divided by two and half, it is little different from the crop in the unfertilized pond (180 kg); the increased crop in the former might easily have been due to the larger number of fish. Where possible, the fish should be of the same origin, preferably the same parentage, and at least the same age. In Malacca each row of Latin square received fish from the same stock, so that each treatment had fish of the same age and origin. A similar arrangement could be used in randomized plots.

The Size of the Experimental Ponds

It has been shown that the growth rate of fish increases with the increase in the area of pond; the "living space" effect has been shown to be a real one. There is also evidence that small ponds may inhibit the growth rate of fish. Ponds should therefore be as large as possible, and all of the same size, except in studies of the effect of size of ponds.

Simplicity of Design

Research workers, especially when they are new to the field, often want to try out complex and sophisticated techniques when an experiment of much simpler design may give the same result with far less effort. It is advisable to start from the simple and work up to the complex rather than the reverse; complexities introduced early are difficult to eliminate later. There are often ways of simplifying an experiment and at the same time eliminating unknown variables. The divided pond technique in Malacca is an example. Ponds divided into one-third and two-third can be used to test the living space effect, while ponds divided down the middle could be used to compare the growth rate of two different strains of fish, as of *Tilapia mossambica*.

When known variables are to be considered, it is wise to limit the number of variables to be studied at one time. In industry it is often possible to have many variables, but the aim of the statistical study is to give greater control, and frequently much is known to suggest the result. In fish ponds there are far too many unknowns, particularly interactions between fish and fish, fish and plankton, soil and water chemistry and pond biota. It is therefore unwise to mix different kinds of variables. Fertilizer studies are better kept separate from stocking rate studies, particularly if several species of fish are involved.

Statistical Designs

The most important variable is fish ponds is that due to uneven soil fertility, slope, drainage, *etc.* There are several statistical designs to reduce the error due to such variability.

Randomized Blocks

In this the total area is divided into compact blocks, each containing the same number of ponds, which should be rectangular and lying side by side. The number of ponds within each block determines the number of replications for each treatment.

Thus, if there are five ponds in each block, five different fertilizer treatments could be tested. The treatments are assigned to the ponds in each block at random, using random numbers. Defects are that the number of replications should be large, not less than five. This means a large area of land must be converted to ponds for this purpose.

The Latin Square

Variations in the soil usually occur in strips or blocks. The error due to these variations can be reduced by using a Latin square design. In this the ponds are arranged in a square pattern, the number on each side being the number of treatments to be tested. Each treatment occurs only once in each horizontal and vertical row, so that errors are reduced in two directions at right angles.

A typical 5 x 5 square is :

```
B   C   E   D   A
E   A   C   B   D
A   B   D   E   C
D   E   A   C   B
C   D   B   A   E
```

It can be mathematically shown that the number of such squares with side n is (factorial n)2. Usually a square is chosen at random from tables, but not all squares are equally suitable, as the following shows :

```
A   B   C   D   E
B   C   D   E   A
C   D   E   A   B
D   E   A   B   C
E   A   B   C   D
```

This is formed by shifting one letter from the end of each row, but it is obvious that treatments lie in diagonals. Any bias in soil in this direction would have a marked effect on the results. In Malacca a 6 x 6 Latin square was chosen.

```
A   B   C   D   E   F
D   F   A   B   C   E
```

```
C   A   E   F   D   B
E   C   F   A   B   D
B   E   D   C   F   A
F   D   B   E   A   C
```

Even here there are too many diagonal groups- four "Fs, three "D"s and three "Bs. Fortunately statistical analysis has shown that there is no bias of the soil in this direction, but it might have been otherwise.

A little shifting of rows would improve the design:

F D A B C E

A C E F D B

C E F A B D

E B D C F A

D F B E A C

B A C D E F

Even this is not perfect and it is difficult to find ideal solutions for a 6 x 6 Latin square. For this and other reasons a 5 x 5 or a 7 x 7 Latin square would have been preferable.

Graeco-Latin Square

Considering two Latin squares in which one can be imposed on the other to form Graeco-Latin square. In this no combination of Latin and Greek letters occurs more than once. Latin squares which can be superimposed on each other are orthogonal to each other. It is possible to have a third, fourth *etc.* squares superimposed in this way, which would permit a different variable for each Latin square. In general if p is a prime number as the side, there are p-1 orthogonal Latin squares and therefore p-1 variables can be tested at the same time. Thus a 5 x 6 square permits 4, while a 7 x 7 square permits 6 variables. Unfortunately, if the side is 4n + 2 orthogonal squares are not possible and only one variable can be tested. Thus it was unfortunate that the designers of the Institute in Malacca chose a 6 x 6 Latin square.

The ponds themselves need not be square, only the pattern of arrangement, but all the ponds must be of the same size and shape and equidistant from one another

Youden Squares

In randomized blocks each block must be large enough to include one of each treatment, but it may happen that one wish to test more treatments than one have ponds in each block. In this case a Youden square is a useful device. If one have blocks of four ponds and one wish to test five treatments one should use the pattern :

	Blocks				
Treatments	A	B	C	D	E
	D	E	A	B	C
	B	C	D	E	A
	E	A	B	C	D

In each block a single treatment has been omitted, a different one in each case. This is in effect a Latin square with the bottom line omitted, and this is an easy way of forming a Youden square. One could leave out more rows; in general if there are p blocks and p-n ponds one can form a Youden square by leaving out the bottom n rows from a p x p Latin square if one wish to test p treatments.

Reduced Latin Squares

The principle of leaving out rows can be extended in two directions and this can make a 6 x 6 Latin square more versatile, regarding it as a 7 x 7 square with one vertical and one horizontal line omitted.

A B C D E F G

B F E G G A D

C D A E B G F

D C G A F E B

E G B F A D C

F A D C G B E

G E F B D C A

This can be extended to a Graeco-Latin square so that two variables could be tested in a 6 x 6 square.

A1 B2 C3 D4 E5 F6 G7

B3 F7 E6 G1 C4 A2 D5

C7 D1 A5 E2 B6 G4 F3

D6 C5 G2 A3 F1 E7 B4

E4 G6 B1 F5 A7 D3 C2

F2 A4 D7 C6 G3 B5 E1

G5 E2 F4 b7 D2 c1 A6

It must be noted that while six of each variables have been omitted twice, that is, horizontally and vertically, one each of A and 6 have been omitted once. For this reason, in the statistical analysis the omitted treatments must be included as null treatment and null crops, when calculating variance.

Collection of Data

Observations during experiments can be of two kinds, descriptive ones, often of great value in giving a general picture, and whose involving measurement. The latter form the data for statistical analysis. Unfortunately, if the numerous research papers on fish culture are any indication, the collection of data frequently leaves very much to be desired.

The Fish

Measurements of fish usually involve the length, depth and weight. Various types of length measurements are used, some to the tip of the tail. Since the tail can be easily damaged, it is preferable to use the standard length, which is the length from the tip of upper jaw to the midpoint of the junction of the fleshy peduncle with the tail. Depth should be the maximum depth, and it is useful to state how far this is along the body line, either in scale numbers or more easily as a proportion of the standard length from the tip of the jaw. In all cases the method of measuring should be stated. Measurement of weight should present no difficulties.

When measuring the total crop of fish, the fish should be harvested after the same period of time after stocking. The growth rate of a fish is not uniform throughout its life, and fingerlings grow much faster than adults, since they have a smaller maintenance requirement. It is not valid to extrapolate growth rates from shorter intervals, as has been done in some research papers. If samples of fish are being examined for comparative growth rates, the intervals of time between fishing should be the same for all ponds and for all fish stocks.

Ideally all the fish in the pond should be measured for growth rates, but this is often not possible. In general not less than 100 of each species from each pond should be measured, for statistical reasons it is preferable that at least one-third is measured; where possible this limit should be maintained. In all experiments containing the same populations the sample size should be the same. When comparing stocking rates the sample should contain the same proportion of the total population, unless this proportion fall below 100, when all should be measured. A similar criterion should be used in the "living-space" experiment. In any case the number of the total population and the number in the sample should be clearly stated. Varying numbers and proportions can be analyzed statistically, but the error mar be greater and the analysis more complex.

Chemical and Physical Measurement

The main object of fish culture is the production of fish and one must never lose sight of this fact. In general limnology detailed surveys are often made of lakes, at various depths and over a period of time. As useful as this is in limnology, in fish culture research one have not the personnel or time for such elaborate studies, nor are they necessarily relevant to the final object, the production of fish. In practice, one or two sampling points are chosen, based on an initial survey of the pond. Having chosen a sampling site and should adhere to this throughout the experiment.

The time of day at which samples are taken is important, particularly if oxygen, pH, alkalinity and temperature are measured since these fluctuate throughout the day. Twenty four hour cycles are of interest, but may not be sufficiently relevant to fish production to warrant the time spent. If data are to be statistically analyzed, they must be sufficiently detailed and ready for analysis as soon as possible. If it is necessary to wait a long period for someone to try and make sense of it, it is hardly worth the effort of collection. Staff is often limited, and with a large number of ponds, this limits the number of samples per pond at any one time. Sunshine, cloudiness, and temperature all affect the results, so if the number of ponds is large there may be appreciable differences between the first and the last, due simply to the lapse of time. The situation is worse if if ponds in different areas are being sampled on the same day. Measurements of pH and oxygen so often quoted in papers may have little meaning because they are taken at the wrong time of day. They may serve as rough comparisons between ponds but cannot be analyzed statistically. Measurements of oxygen just before dawn, when the values are a minimum, have value for indicating danger levels to fish.

Research papers often list estimates of dissolved phosphate, potassium and nitrogen, but it is reasonable to ask whether such figures have any real meaning. They may have value in large lakes, or in unfertilized smaller bodies of water, but in fish ponds where fertilizer treatments can radically alter conditions, such isolated measurements have little value. Phosphate rapidly disappears from the water, being absorbed by phytoplankton, macrophytes and the bottom mud. There will therefore be a marked difference between estimates immediately after fertilization and at a later date. In fertilizer trials the chemical analysis of water (and if possible plankton, macrophytes and mud) should be planned as a complete and integral part of the program and the methods should be standardized – sampling site, time of day, method of sampling and methods of estimation. The chemical analysis should be undertaken by a chemist specifically for this work. It still does not necessarily follow that the chemical analysis will show any clear relation with the fish crop.

Plankton

Although it is customary to take phytoplankton and zooplankton samples throughout fish culture experiments, it is still necessary to be clear what any estimates if any actually mean. Phytoplankton in fish ponds is dense, often forming scums. Wind may pile up these algal scums, to one end, particularly, *Microcystis, Euglena etc.* Cyanophyceae with gas vacuoles, and mucilaginous green algae which can trap oxygen, may rise during the day and sink at night. Since time will not permit sampling at several points the sampling points should be at the end away from the algal scums. An accurate estimate cannot be obtained, but a picture of the relative change with time is possible.

Zooplankton often shows diurnal vertical movements, and much which applies to phytoplankton applies to zooplankton also. The sampling points, time of sampling and the intervals of sampling should be fixed. Methods of estimating the zooplankton should be consistent, and if it is necessary to dilute for counting, the dilution should be stated. Units, whether as single cells, filaments or colonies should be stated, and these units should be consistent.

With small samples from large populations, variations of as much as 100 percent may occur through sampling error. It would therefore be unwise to regard an increase from say 500 to 900 organisms per ml as being significant, without other evidence to support it. In general, changes within the same logarithmic paper should be regarded as not significant. A change from 500 to 5000 could be regarded as significant.

Extrapolations from plankton measurements to the annual production are of doubtful value. This is particularly so in fish ponds, where the high grazing intensity of the fish markedly affects the standing crop, so that one have no measure of the true primary production. Although pictures of the relative changes can sometimes be deduced, it would be unwise to attempt the analysis of such data statistically. Primary production can be measured reasonably accurately by using dark and light bottles or C14, but this is total primary production. Since the ability of different algae to be digested by fish varies, one have no measure of how much of the total primary production is utilized by the fish. Plankton counts of genera and species, together with a study of which species are digested may give a general picture of the production of the pond, but such data cannot be analyzed statistically.

Presentation and Interpretation of Data

Proper presentation of data is of special importance in fish culture research. Fantastic annual fish crops are often claimed, but on close examination these appear to be extrapolation of growth rates of small fish, hardly bigger than fingerlings, taken over a short period of time. The growth rate of a fish is not a straight line graph which can be extrapolated in this way. The maintenance ration for small fish is low, so that most food consumed goes into growth, whereas the maintenance ration of large fish is much greater. It is possible to get a larger crop of small fish than of fish of edible market size, but the proportion of utilizable material may be greater in the larger fish. Whenever crops of fish are the object of the experiment, it should be extended until the fish reach edible size. In some places this is often six months in growth, but elsewhere it may be longer. Standard and consistent measurements should be used and stated. The size of pond should always be stated as well as the stocking rate.

Chemical data presented should be relevant to the experiment. There is little point in making long lists of chemical analysis and repeating these in words, when they have little connection with the experiment. Data should be concise and to the point.

This applies equally to plankton. Long lists of species and genera often have little relevance to the experiment. Significant fluctuations are better treated graphically and the attention drawn to the fluctuations, where possible with conclusions. It should be relevant to the treatment of pond or some important feature in the experiment. It may not be possible to treat these data statistically, but if given with other relevant data, such as the feeding habits of the fish, they may enable interpretation of data statistically.

If many of the points (on the collection of data, with various pitfalls) seem obvious, judging from many research papers they may be so obvious that they are neglected by many research workers.

Chapter 2

Evaluation and Assessment of Aquatic Communities in a Water Body

Method of Collection and Evaluation

Bacteria

In natural ecosystems, the microbial communities have adapted to their environment and the species composition of given habitat appear to be characteristic. Differences between undisturbed and disturbed environments can be detected, not only by alteration of the species composition, but also by changes in heterotrophic potential, total biomass as estimated by epifluorescence or ATP measurement and other parameters.

Microbial indicators should, therefore be developed for estimation of the environmental implicacy at the microbial level. It is useful to emphasize that the microbial response is, in general, rapid, that is, within hours of impact, whereas higher organisms may not demonstrate effects for weeks or months, except of course, as a result of massive influx of toxic effluent.

Relatively undocumented are the effects of multiple pollutant input, as for example, domestic sewage and acid plant water wastes entering in the same stream. The effects of the acid plant water are, most frequently, bactericidal with marked reduction in bacterial numbers because of the low pH induced in the receiving streams. Should the acid plant wastes be neutralized, the sewage effluent, thereby, becomes a serious public health hazard, if the wastes are not appropriately treated prior to discharge. Thus, the microbial effects of environmental impact must be properly assessed in terms of the total pollutant discharge and their effects on the ecosystem.

Sampling Methods

Collecting samples for microbiological analysis require aseptic technique, including a sterile collecting device. Methods for collecting water and sediment samples for microbiological analysis have been reviewed (Colwell, 1974). Several sterile water samplers, including the J-Z and the Niskin samplers, or modifications of these, are available commercially. Unfortunately, sampling devices for aseptic sediment are not available. Coring devices or grab samplers are usually employed for collection of sediment for microbiological analysis.

Samples must be collected at given times and in replicate in order to achieve statistical validity.

Microbiological Analysis

Environmental impact at the microbial level is best achieved by enumeration of the total biomass, estimation of heterotrophic capacity for substrate uptake, and determination of selected "indicator species".

Water and sediment samples collected from waters receiving acid plant wastes should be analyzed for total biomass, using methods, such as, ATP, epifluorescence, Limulus lysate (LPS), or muramic acid assay. Total viable aerobic heterotrophic counts are also helpful, since total numbers of aerobic, heterotrophic bacteria can provide an index useful in assessing the quality of the receiving waters and/or the effects of the acid plant waste discharge. Most probable number estimates of *Thiobacillus* spp. could provide a useful index, as well, since the numbers of these and of acid-tolerant bacteria should rise with increased impact from acid plant drainage.

In waters receiving untreated domestic wastes, as well as acid plant drainage, it is imperative that monitoring of bacteria of public health significance be done, particularly if the acid plant waste discharge is erratic and pH shifts occur. The die-off of coliforms in acidic waters would fluctuate, depending upon the concentration of nutrients in the discharge, as well as fluctuations, if any, in pH.

Metal-resistant bacteria should be assayed, since populations of metal-resistant bacteria have been found to be correlated with concentrations of metals in the aquatic system. Furthermore, heavy metal resistant bacteria have been found to be markedly resistant to antibiotics. Thus, in areas receiving domestic wastes and acid plant drainage, there is the possibility of selection for acid-tolerant, metal-resistant sewage bacteria that may also be antibiotic resistant. The potential public health risks of such situations should not be minimized.

Phytoplankton

Phytoplankton are small plants, mostly microscopic, that either are weakly motile or drift in water, subject to the action of waves and current, consisting principally of the free living bacteria and algae. The phytoplankton form the base of the food web in many habitats and, therefore, may be extremely important to the economy of the ecosystem. The principal properties of the phytoplankton related to the assessment of the adverse impact of pollutant on aquatic system are as follows;

1. Standing crop (numbers and/or biomass per unit sample volume);
   (a) Count (cells or clumps)
   (b) Cell volume,
   (c) Ash-free weight
   (d) Cholorophyll *a*
   (e) ATP

2. Community structure
   Species composition
   (i) Indicator species,
   (ii) Number of individuals per species,
   (iii) Species richness,
   (iv) Diversity index (*d*)

   Pigment composition
   (i) Autotrophic index (AFW/chl *a*)
   (ii) Chlorophyll *a*/Chlorophyll *b* ratio,
   (iii) Chlorophyll *a*/Chlorophyll *c* ratio

   Nitrogen fixation

   Metabolic activity – Primary productivity
   (i) Carbon – 14 uptake
   (ii) Oxygen evaluation

   Respiration rate
   (i) Dark bottle oxygen uptake
   (ii) Electron transport.

The taxonomic composition of phytoplankton communities generally follows pronounced annual season cycles and the standing crop may vary greatly over short period of time. The phytoplankton are generally not important in small streams (where the flow is less than one cubic meter per second), and should not be included in assessment programs.

Because of the temporal and spatial variability in phytoplankton distribution samples should be collected frequently (at least weekly) and in replicate (3 replicate minimum).

Sample Collection and Preservation

Samples may be collected with water bottles or by pump. The use of nets should be avoided.

In flowing water, samples are taken mid-stream near the surface. In standing water (reservoirs, lakes, estuaries and coastal water) samples are collected in the epilimnion, near surface, mid-depth, and at the metalimnion.

Sample Analysis

Counting and Identification

Organisms are counted and identified by using the Sedgwick-Rafter or inverted microscope method. Quality assurance must be maintained by use of a stage micrometer and other calibration devices.

Biomass

The dry weight of particulate organic matter (ash-free weight) concentrated from surface waters is a useful measure of phytoplankton biomass. If large numbers of zooplankton, bacteria and/or organic detritus are present; however, the data may not be representative of the algal biomass.

Chlorophyll *a* is a specific index of phytoplankton biomass, and data on chlorophyll *b* and *c* provides information on the taxonomic composition of the algae. Chloprphyll *a* may be determined by spectrophotometric or flurometric methods, and should be corrected for the presence of pheopigments.

The Adenosine Triphosphate content of the plankton is an index of viable biomass, and may be useful in measuring near and long term effects on the standing crop and condition of the phytoplankton. It is a nonspecific microbial biomass index; however, and may be difficult to relate to algal standing crops if large numbers of bacteria or other non-cholorophyllous microbes are present.

Quality assurance is maintained in biomass measurements by the use of appropriate standard weights, spectrophotometric and fluorometric calibration materials, and the use of standard biological reference samples.

Metabolic Activity

Rates of metabolic processes, such as, photosynthesis, respiration and nitrogen fixation, are useful in measuring the environmental impact of pollutants on receiving waters. Photosynthesis and respiration rates may be depressed or stimulated depending upon whether the pollutant is toxic, bio-stimulatory, or otherwise affects the chemical or physical properties of the aquatic environment. Nitrogen fixation rates are related to the taxonomic composition of the plankton and indicate trends in community structure, Data on nitrogen fixation in surface waters are still very limited, and the measurement of this parameter is of low priority.

Photosynthesis Rates

Photosynthesis is commonly measured by the Carbon – 14 uptake or oxygen evaluation method using samples enclosed in "light" and "dark" bottles. The number of depth intervals chosen must be sufficient to define the photosynthesis curve. An interval of $1/10$ of the photic zone is recommended.

Nitrogen Fixation

Nitrogen fixation rates are measured using the acetylene reduction method. The ethylene produced is quantified by gas chromatography.

Data Evaluation and Utilization

Data on standing crop, community structure and metabolic rates in the polluted waters are compared to conditions at the control (unaffected) stations.

Data on species composition can be used to determine the effects of the pollutants on community structure. Numerical species diversity indices, such as *d* (Wihlm, 1970) can also be used to compare the structure of phytoplankton communities in the control and affected areas. Numerical indices must be used with caution, however, because of the potential confounding effects of other factors which may not be immediately apparent.

Macrophyton

Definition

The term "aquatic macrophyte" refers to larger aquatic plants, as distinct from microscopic algae. Macrophytes possess a multicellular structure, usually with cells differentiated into tissues. Mycrophyte encompass a large assemblage of plants, such as, macroalgae, mosses, liverworts, ferns, and vascular plants ranging in size from the near-microscopic watermeal to massive cypress trees.

Macrophytes are useful in assessing chemical and physical perturbations to the environment. Macrophytes are usually fixed at a location and are sensitive to many types of stress, including thermal stress inputs which influence water clarity (dredging, mining, industrial discharge), substrate changes (channelization, sediment deposition), organic pollution, nutrient enrichment, and toxic materials such as herbicides. Because of their ability to assimilate and retain contaminants, macrophytes are useful for detecting discharges from intermittent sources that may be missed in routine chemical studies.

Importance

Aquatic macrophytes are important structural and functional components of many marine and freshwater ecosystems, and they influence these systems in a variety of ways. Macrophytes function beneficially in aquatic systems to :

☆ Convert light energy and mineral nutrients to organic matter.

☆ Serve as an important consumer food source through grazing and detrital food webs.

☆ Provide substrate for a diverse assemblage of attached microscopic and macroscopic plants and animals.

☆ Provide spawning substrate for aquatic vertebrates and invertebrates.

☆ Serve as a cover and nursery area for both aquatic vertebrates and invertebrates.

☆ Trap and recycle nutrients.

☆ Build and stabilize substrate. Serve as a source of oxygen.

Excessive growths of aquatic macrophytes may create a nuisance by impeding navigation, eliminating spawning habitat, interfering with fishing, and causing severe diurnal fluctuations in oxygen, often with total depletion of oxygen during hours of darkness. Mechanical cutting, application of herbicides, and habitat alteration are the primary control methods.

Information about macrophytes is useful when initiation and evaluating control programs. The primary aim is to control plants that are nuisances and to leave areas in a natural state that do not interfere with varied uses. It is important, therefore, to understand the ecology of aquatic plants and to use them to the best advantage in maintaining aquatic systems in a natural state where they are useful for fish and wildlife production and recreational uses.

Aquatic macrophyte communities frequently exhibit both vertical and horizontal zonation and are conveniently divided into three categories based on growth form and zone of habitation.

☆ *Floating* – This category includes plant not anchored to the bottom such as, duckweed, water hyacinth, and some macroalgae which float on the water surface. They are subject to winds, tides and currents, so they are normally found in abundance only in sheltered areas.

☆ *Submergent* – Plants anchored to the bottom by roots, rhizomes, or holdfasts. They may be entirely submerged or may have floating and aerial reproductive structures (namely, water milfoil, eel grass, bladderwort). Submergent plants are found from the shoreline to a depth determined by water clarity, substrate and growth form.

☆ *Emergent* – Plants occurring on soils saturated with water or on soils covered with water most of the growing season. The two major groups are floating-leafed plants (water lilies) and plants with upright shoots (cattails, wild rice, sedges, woody shrubs and trees).

Assessment

There are numerous parameters useful in evaluating the structure and function of aquatic macrophyte communities. Parameter selection is depended upon study objectives, time and resources available. Parameters of community structure include :

☆ Species present

☆ Percent species composition

☆ Diversity and equitability.

☆ Arial (vertical and horizontal) zonation.

☆ Temporal distribution.

Parameters of community function include :

☆ Nutrient and metals uptake and cycling.

☆ Productivity – Production of biomass, total standing crop, biomass turnover, stem counts, arial coverage and detrital import and export.

☆ Oxygen evolution and uptake.

☆ Growth and death-rate measurements.

Sampling and Analysis of Macrophytes

Knowledge of the kinds of plants present in an area, percentage species composition, diversity, standing crop, arial coverage, arial zonation and temporal distribution are basic parameters for describing community structure and determining environmental perturbations on surveys.

In many cases, environmental perturbations to not have an immediately observable effect on plants but remain latent for the life span of extant vegetation. Effects are often observed in growth rates, longevity and vitality of the aquatic flora, and in colonization patterns of new substrates or replacement of older communities. Such effects can not be easily assessed with traditional methods and studies of short duration, so changes in longevity, mortality and appearance of extant vegetation, measurement of recruitment and establishment of new vegetation and determination of in situ effects of environmental perturbations on particular species of plants are other methods used in macrophyte studies.

Field Sampling Methodology and Equipment

Regional Rurvey

Regional floristic surveys are still commonly reported because they are relatively easy to do and distribution patterns of many plants are poorly documented. Sampling includes visual observation and collection of representative types from the study area and delineating plant associations. Sampling gear is varied, and the choice of tools usually depends on water depth. Dredging or SCUBA diving are often employed to obtain deeper samples. Regional surveys are usually restricted to qualitative observations which results in reports of species lists and floral assemblages.

Line Transect

Line transects are frequently employed in an attempt to obtain quantitative information. A weighted nylon line or lead core, usually marked off in meters, is laid out along a compass line to obtain the transect. In deeper areas, the line is laid from the surface or by divers directly on the bottom. Numbers of plants occurring along the line or the linear distance occupied by each plant along the line is recorded. Data are often expressed in profile or life form association diagrams. This type of transect is tedious and time consuming to conduct, especially in under water surveys. There are also parallel problems in deciding what lies under a line and, in the case of foliose algae, how much of a frond should be counted.

Belt Transect

A belt transect is similar to a line transect, but instead of a line, a belt of vegetation is surveyed. One advantage of the belt transect is that it can be treated as a series of independent continuous quadrants adjacent to one another, forming a strip. Each "quadrant" is defined either as the space within a marked region or the entire field photographed from a standard height. Photography provides a photographic record that can be analyzed in detail later. For deep water surveys, a single transect line is employed in conjunction with one or more weighted bars used to determine the transect width. All plants crossed by the bar while swimming along the transect line are counted. In the photographimetric method, bars or marking on the transect rope are used to delineate the edge of each photo-field. Similar methods are used in shallow waters. Data are usually recorded as numbers of individuals or percent cover of each species.

Quadrant

One of the most widely applicable techniques is that of sampling within a quadrant or plot of standard size. Quadrant sampling techniques may be adapted for use in all major types of plant communities. Quadrants can be sampled more rapidly and are much easier to work with than either line or belt transects.

The details of quadrant sampling procedures including size, shape, and number and arrangement of the sample plots must be determined for the particular type of community being sampled on the basis of the type of information desired.

For sampling in shallow water, large areas can be outlined with stakes and cord. Smaller areas can be encompassed with wood or metal frames made into four pieces that slide horizontally into the vegetation and are fastened together to form a square. Another useful sampling device for shallow waters is the stovepipe which consists of a light metal tube or frame with two handles for pushing the tube into the substrate. A decision must be made whether to include only plants rooted within a quadrant or to sample all plant material with an upward projection irregardless of where the roots are located. This decision is regulated by the type of plant, time of year, density of growth, type of habitat and other considerations. Actual quadrant locations are either determined by blind casting a marker buoy or by selection of a particular occasion along a transect line. When quadrant sizes are relatively small, a nested quadrant technique is useful. In this technique a large quadrant is positioned and then a smaller quadrant within the larger one is chosen in a random manner for analysis.

In deeper water where visibility is poor and not conducive to SCUBA diving, a number of epibenthic samplers such as the Peterson and Ponar grabs, weighted oyster tongs, *etc.* can be used. There are problems associated with these sampling devices because (a) they are operated blindly from the surface, (b) they are cumbersome and slow, (c) they rely on mechanical closing devices that often fail, (d) they operate poorly on hard sand or shell bottoms, (e) they remove roots and rhizomes inadequately, and (f) are prone to errors arising from the wrongful inclusion or exclusion of weeds at the edges. If they have to be used, one suited to the type of plant and bottom should be selected. Purkerson and Davis (1975) have developed and evaluated a quantitative epibenthic sampler that solves most of the above problems for shallow (less than 30 meter deep) areas with underwater visibility greater than 80 centimeters. The P and D sampler takes advantage of recent advances in SCUBA to permit in situ quantitative collection of plant material. It consists of a clear plastic box and a mesh bag. The outside dimension of the box are : 45 cm (length) X 45 cm (width) X 34.5 cm (height). A 3.5 cm frame of angle iron is attached to the bottom of the box to control sample depth. The box is positioned by a SCUBA diver and thrust into the substrate until the horizontal metal flange is flushed with the ambient consolidated sediment. Working in glove box fashion through the opening adjacent to the mesh bag, a SCUBA diver removes all the surface biota (grasses, macroalgae, alcyonarians, sponges, *etc.*) with garden shears or by hand. Plants are transferred to the bag with sweeping motions that create water flow in through the hand opening and out through the bag, preventing loss of material. After all surface material is in the bag, the top 3.5 to 4 centimeters of the sediment is transferred to the bag by hand in the same fashion. The vertical metal flange serves as a guide for sample depth.

A sampling device developed by Gale (1974) utilizes SCUBA techniques to collect benthic samples. The 18-inch diameter Suction Dome Sampler has a serrated stainless steel band along the lower edge for penetration into sand, pebble, gravel and/or cobble substrates. The top consists of anacrylic hemisphere. Organisms are pumped from inside the sampler by a venture pump (powered by a 12-volt motorcycle battery) into a 14 x 36 inch monofilament collection bag. The Dome sampler can be used in lentic and lotic waters. A lead belt inside helps hold it in place in strong currents. Samples can be collectd by wading snorkeling in water up to 3.0 feet deep and in deeper water by SCUBA diving.

In riverine system where there is adequate current, macrophytes can be collected with Surber sampler or they can be removed from a quadrant and collected in a net below the sampling area. Sampling device usually enclose an area from 0.04 to 0.5 square meters. Size of the device used depends upon vegetation density. Where blades or fronds are dense (2000 per square meter or more), a 0.04-square meter area is a more practical area to use. Plants rooted in soft mud or silt can be pulled out by hand. Plants in other substrates or plants with massive rhizome systems must be dug out or pulled out by a heavy sampling device such as oyster tongs.

Leaf Area

The photosynthetic surface area is measured by obtaining a two dimensional outline of the frond. One of the best ways to do this is to make Xerox or similar reproduction of the frond and determine the area with a planimeter. Photographimetric methods are easiest to work with if material cannot be collected; however camera angle and distance to subject must be standardized as much as possible. Leaf area data can be utilized to estimate habitat space and photosynthetic efficiency when combined with productivity data (Odum et. al., 1963) It can also be used to determine whether density and light are controlling maximum standing crop. The leaf area index (LAI) is an estimate of this maximum leaf density. In terrestrial cereal crops, maximum values of the LAI are as high as 9; for broad leaved trees in a tropical rain forest it can read 20. In sea grasses, LAI over 20 can be attained in a dense meadow. This very high values suggest some adaptation to the lower light levels of a submarine life; it is also possible that the highest LAI for any sea grass species can only be reached where some degree of water movement is present. The consequence of increasing the bottom area by more than 20 times is, of course, extremely important to other inhabitants of the community (Mc Roy and McMillan, 1977).

Meristic Measurements

Detailed meristic measurements, Such as, stipe length, frond width, fruiting and vegetative reproduction can be compared in potentially stressed areas and uninfluenced adjacent areas to distinguish effects of perturbations on macrophyte populations.

Treatment of Collected Plants

Macrophytes are usually processed either while wet or after drying. Samples collected in the field can be identified, separated immediately, placed in plastic bags, and refrigerated; or pressed and dried with a plant press for identification and processing. If a time lapse of more than two weeks occurs between sample collection and laboratory processing, then plants should be preserved in 4-percent buffered formalin or dried.

Reasonable distinction between live and dead plants or plant parts can be satisfactorily determined through visual examination, chlorophyll *a* analysis, and or vital staining techniques.

Gaff and Okong' O-Ogola (1971) developed a stain technique which is readily adaptable to field use. Using Evan's blue (Chroma), a non-toxic water-soluble dye, they developed a rapid staining method for distinguishing dead from live plant tissue. Evan's blue has the inverse staining properties of a vital stain, and the staining test depends basically on the retention or loss of the semi-permeable properties of the plasma lemma. Living cells or tissues exclude Evan's blue, whereas the dye penetrates and stains dead cells.

The Evan's blue technique has been applied successfully to a large number of species from a variety of families. It is applicable to non-vacuolated as well as to vacuolated cells. Since the test solution is excluded from the living cells, it is unlikely to induce changes in cells if metabolic studies are to be conducted, consequently, it may be applied repeatedly to the same cells.

In the laboratory, plants should be separated and cleaned free from debris and periphyton by hand. Once plants are cleaned of adherent materials they should be identified and preserved.

Total biomass of large samples can be determined by spreading the plant mass on large trays and drying at 105 degree Celsius for 24 hours or longer and then grinding the plants in a blender until uniform particle sizes are attained. Samples then can be conveniently processed for biomass determinations or ground samples can be sub-sampled for biomass or other chemical determinations.

Dry weight is determined by drying representative samples for24 hours or to a constant weight in an oven at 106 degree Celsius. After samples have dried for a specified time period, they should be cooled for 1 hour in a dessicator before dry weights are determined on accurate scales.

To ascertain the organic content of a sample, incinerate it in a muffle furnace at 550 degree Celsius for 1 to 6 hours, depending on the amount of material to be ashed; cool the ashed sample in a dessicator; rewet it; and dry it for 24 hours at 105 degree Celsius. Removed the ashed sample from the oven, place it in a dessicator, and allow it to cool for 1 hour; then weight it to obtain ash content. Determine organic matter by subtracting the ash content (ash weight) from the dry weight and calculate the grams per square meter of dry weight organic matter (ash-free weight).

Overestimation of dry weight and under-estimation of organic matter content can occur when plants are encrusted with a considerable amount of carbonate. To eliminate these carbonate encrustations, samples should be placed in a 10-percent hydrochloric acid solution for 15 minutes or more. Treatment time depends on the amount of encrusted material on plants. Complete dissolution of carbonate material from the plants can be determined by visual or microscopic inspection. Upon dissolution of the carbonates, samples should be washed at least six times with distilled water to remove hydrochloric acid. Pour the last washing into a vessel and add a few drops of silver nitrate. If a white precipitate (silver chloride) does not form, then the sample is free of chloride ions which could affect weight analysis, and it can be dried.

Dried samples for chemical analysis should be done according to EPA Chemical Methods (1974) or Food and Drug Methods (1971).

Diversity and Equitability

Two important components of community structure are species richness and distribution of individuals among the species. These components can be summarized into diversity indices, which equate the amount of uncertainty that exists regarding the species of an individual selected at random from a community. The more species there are and the more nearly even their representation, the greater the uncertainty and, hence, the greater the diversity. While it is inappropriate to apply the evenness concept to complex communities having several trophic levels, the concept is applicable to single-trophic level communities such as vascular plants and macro-algae.

Odum (1971) stated that the pyramid of numbers is not very fundamental or instructive as an illustrative device because of the geometrically large unit. The pyramid of biomass is of more fundamental interest, since the geometric factor is eliminated and the quantitative relations of the standing crop are well shown. In the mathematical expression of data in diversity indices, usually numbers of individuals in the different species have been used with no consideration of variation in weight among species. With certain organisms, such as, diatoms, weight variation may be slight and insignificant; however, differences in weight may be considerable among populations of macroalgae and some vascular plants. This discrepancy can be corrected by using biomass units instead of individuals which are very difficult to discern, especially in macroalgae and some vascular plants having unique means of vegetative reproduction.

The modification from numbers to biomass units (mg/square meter) necessitates substituting a continuous variable for a discrete variable, but it redefines diversity in biomass terms. Diversity is equated with uncertainty regarding biomass instead of numbers. The more species there are and the more nearly even their biomass representation, the greater the uncertainty and the greater the diversity (Wlhm, 1968).

Aufwuchs

Usually the term "periphyton" refers to the community of organisms actually attached to submerged aquatic substrate, either natural or artificial. The term "phycoperiphyton"(Foerster and Schlichting, 1965) for attached plants and "Zooperiphyton" (Cairns, 1978) for attached animals have also been used. The term "aufwuchs" (a German term without English equivalent) is sometimes used as a synonym of periphyton. This is unfortunate, since the distinction is operationally very important as different methods are needed to sample the aufwuchs and its sub-community, the periphyton. Ruttner (1953) gives the following definition. By the term aufwuchs, we mean all those organisms that are firmly attached to a substrate but do not penetrate it."In addition to attached species there belong to the aufwuchs biocoenose a large number of free-living forms, which crawl upon the substrate, swim about in the dense confusion of the sessile species, or even undertake further temporary excursions into open water "The aufwuchs clearly require more care to collect, study and count than the periphyton, which is firmly attached and will remain intact on natural or artificial substrate when they are lifted directly out of the water. In contrast, the aufwuchs and its substrate must be completely enclosed in a container to be collected undisturbed. It is the aufwuchs community that is the functional whole while the phycoperiphyton, zooperiphyton or any other segment is only a fragment of the aufwuchs community. In describing the aufwuchs community it can be called as the "slime coat" community which is possibly more meaningful than "aufwuchs".

The aufwuchs community is the most trophically complete of the various communities – plankton (floaters, small swimmers), benthos (burrowers) and nekton (large, strong swimmers,) and osmotrophic decomposers, detritivores, scavengers, herbivores, carnivores, as well as chemosynthetic and photosynthetic autrophs. Usually the aufwuchs is considered a microbial community with most species having a cross-sectional area less than one mm.The larger animals and macrophytes attached to surfaces are studied by different methods, often not requiring the microscope for identification. In addition to its largest species of protozoa, metazoan, and metaphyta; the aufwuchs community has smaller protistan eukaryotic protozoa, fungi, and algae, as well as the smallest world of prokaryotes, such as, bacteria, spirochaetes, and blue-green algae and tiny eukaryote, amoeba, flagellates and fungi. These three size ranges require different microscopic magnifications, such as, 10-30X, 50-200X,400-1000X respectively, for their identification and enumeration. Also these three size ranges require different sampling methods to obtain reproducible counts.

Methods of Study

To study the aufwuchs, a natural or artificial substrata is required. The natural surface may be living like a plant, or produced by a plant (wood) or animal (shell). Unless the natural substrata is transparent or translucent, epilumination is required to directly study the aufwuchs, and epilumination is usually not very useful in identifying aufwuchs species. Non-living substrata (rock, gravel or sand) are all opaque requiring epilumination for direct observation. One solution is to scrape off the aufwuchs and use standard method counting chambers for observing and counting the organisms Yet a careful comparison of the aufwuchs before and after scraping clearly shows the dispersal, destruction, and death of certain species. Also scraping produces particles that entrap and hide organisms from view. There is a real need for a method that can disperse scraping particles, compress the particles, and also inactivate without harming the motile species, which tend to cluster in standard counting chambers. This need has been fulfilled by combining two separate methods Aufwuchs scrapings (and just as readily, plankton samples, activated sludge or benthos flocculent organic material) placed on a plastic slide are dispersed by mixing a known volume (say 0.02 ml) with a drop of one percent Polyox solution and the drop covered with a thin flexible plastic cover slip cut from Handi-Warp with a grid of about 700 microns marked in the plastic. The non-toxic viscous resin holds the organisms immobile while the thin flexible plastic cover slip compresses the particles and allows one to count the organisms in the whole volume or any portion of the volume using the appropriate sampling method. The very thin plastic cover slip allows abundant aeration during the identification and counting. The catch is that Dow Chemical no longer makes the grided Handiwarp. Plastic film with grid squares to use at the three size ranges of magnification may be used for studying aufwuchs organisms. The thin plastic film cover slips allow for excellent optical qualities.

In recreating the geographic contours of the natural substrate, yet having translucent or nearly transparent surface for translumination microscopy, then the natural surface can be coated with some film such as parlodion which can be peeled off for study. Such methods are especially useful in studying the smallest members of the aufwuchs, since the piece of thin film can be placed under a glass cover slip and studied at highest magnification.

Most of the aufwuchsologists use artificial substrata to collect their charges. Sladeckova in 1962 reviewed the previous literatures and methods, especially the use of glass slides and cover slips. Based on her statements about the limitations of such methods, plastic Petri dish method was developed.

Glass and plastic tend to simulate very different kinds of natural substrate. Glass is very hard, dense, smooth and chemically wettable due to its charged surface. It contains impurities and varies in what it collects depending on how it has been cleaned. Plastic is close to the density of water (requiring little floatation), is softed with mold grooves making it less smooth than glass and more like a plant surface. Also like many plant surfaces plastic is non-wettable with a non polar surface. Aufwuchs organisms attach more rapidly and more firmly to plastic. In the plastic Petri dish method the bottom of the dish with its four projections is pressed into a section of a Styrofoam cup and the bottom is allowed to colonize. Then the tight-fitting lid is applied, excluding all air, and the outside of the bottom portion cleaned. The dish is then placed on a plastic square with an engraved grid of appropriate size and the counting made directly by scanning across the dish. The enclosed aufwuchs community can be studied at up to 100X magnification looking through the clean lid or at up to 440X through the inverted dish bottom. The plastic is less optically clear than glass cover slips and the plastic is so thick (1 mm) that the working distance restricts one to below 440X magnification. Currently these limitations were partially overcome by also collecting with glass cover slips to make the identifications of the smaller species which then can be counted through the less optically clear plastic. What is needed is a plastic Petri dish with tight fitting lid that is especially designed for collecting the aufwuchs with a thinner, more optically clear bottom.

John Cairns has introduced a very useful artificial substrata, polyurethane foam, which has a real advantage for sampling certain bodies of water, however, it does not sample the natural aufwuchs community in its many inner-spaces from which the organisms are expulsed by squeezing the foam. The method has been used extensively and from area to area gives good comparative results.

The most limiting of all factors on the study of the organisms of the aufwuchs community is that the lack of broad training or interest in identifying and enumeration of the whole aufwuchs community. The best known algal group of the aufwuchs is probably the diatom which can be isolated by digesting away all but their siliceous frustules, which can be precisely identified as well as stored for future reference. Yet in some aquatic environments diatoms are absent. Another well-known and highly useful indicator group which can be identified to some degree, is the sessile ciliates, especially the peritrichs and suctorians. They could become as useful as the diatoms in diagnosing aquatic health. They are abundant in turbid, poly-saprobic waters where diatoms are some times lacking.

Zooplankton

Zooplankton are those organisms within the aquatic ecosystem whose distribution and dispersal are influenced significantly by the mixing processes of the water column. Zooplankton are essentially intermediaries in the food chain in that they frequently graze either on phytoplankton or on the detritus with its heterotrophic community, typical in many estuaries, they are in turn used as a food source for higher trophic level predators.

A majority of the animal phyla are represented at some life stage within the zooplankton community. Those plankton which remain a member of zooplankton community throughout their life (holoplankton) are dominated by crustaceans whereas the greater diversity of phylum representation comes from those forms which are a member of the plankton community during only one phase of their life cycle, (meroplanktonic) usually in early life stage. This latter group would include forms such as molluscan larvae, crustacean larvae as well as fish eggs and fish larvae.

The zooplankton community residing in coastal marine waters are typically comprised of copepods, mysid shrimp, many crustacean larvae including cirriped larvae and decapod larvae. Although appearing seasonally, jellyfish and ctenophores are very abundant and are certainly formidable force in zooplankton community dynamics. In an estuary the number of species which are represented is drastically reduced due to the physiological stress imposed by the wide fluctuations in physio-chemical parameters. Again crustaceans copepods are the dominant organisms particularly in the highest salinity areas of the estuary with cladocerans increasing in importance where the salinity gradients to near zero.Other organisms which are prevalent seasonally include the jellyfish and ctenophores as well as annelid larvae. In terms of total numbers in the less saline portion of the tidal river, rotifers are very abundant and serve as the dominant food item particularly for copepods. Also in the freshwater tidal rivers greater numbers of immature aquatic insects are particularly encountered.

In slow moving (lentic) rivers which are punctuated with pools, densities of copepods, cladocerans and rotifers will be present. However, in those rivers or streams with high velocities (lotic) the zooplankton would be very sparse and limited just to those forms which have been recruited upstreams from lakes, ponds or the slow moving portions of the river. In these fast moving waters, the plankton is primarily drifting immature aquatic insects and tychoplankton- those members of the aufwuchs or periphyton which have broken loose.

The zooplankton community of the lakes is comprised primarily of copepods, cladocerans and rotifers.

It is precisely the great diversity of animal types which are present in the zooplankton community which makes this group so difficult to assess quantitatively. Therefore, the collecting methodologies for a particular study must reflect those organisms residing in the biotope to be sampled.

Significance

Data on the zooplankton community composition, distribution, diversity, density and an assessment of the physiological condition through a live : dead ratio have application of the following circumstances :

Community Density

To establish the presence or absence of community alteration resulting from perturbations when compared with reference areas of the community density.

Taxonomic Composition

(a) Spatial

(b) Vertical

(c) Physiological condition – egg size analysis, lipid:carbohydrate ratio, live:dead ratio (neutral red Tetrazolium)

To monitor long term trends in water quality.

To detect the effect of pollutants on indegeneous zooplankton communities in receiving waters.

To determine the effect of entrainment on zooplankton communities (where applicable).

To determine the alteration of age structure of the zooplankton community in a study area when compared to a reference site (applicable primarily to the micro-zooplankton).

Sampling Design

For a quantitative assessment of an impact on the zooplankton community, experience has indicated in all aquatic ecosystems that large volumes of water be filtered in order to integrate the aggregates or "patches" of organisms which are typical of zooplankton distribution on both a vertical and horizontal scale. It is conceivable for a small volume of sample to be collected in a patch that is 2 to 5 times greater in density than found in the background densities. Since primary concern is to be concentrated on large-scale differences, the samples should represent composite "patches".

The vertical distribution is altered by a behavior pattern of vertical migration noted in several of the zooplankton populations whereby the organism will migrate a considerable distance vertically at certain periods of the day. Since the typical pattern of the migration results in the community being distributed higher in the water column at night it is advisable, particularly in the shallow areas, to sample the community at night as some of the larger zooplankton are known to seek depressions along the bottom of shallow rivers and bays seemingly to avoid the intense surface illumination.

Design

In rivers and small river estuaries, the design should include a series of equally spaced collections along a longitudinal transect. These collections can be grouped according to areas for subsequent analysis. These data can be subjected to testing using ANOVA to determine if differences in the areas are significant or other tests including curve fitting might be appropriate.

The same type of analysis would be appropriate in lakes, larger estuarine systems and coastal waters where the samples should be collected in a grid design, sampling all points in the grid or randomly selecting points within a specific time scale.

It is conceivable to design serial sampling along a transect consisting of multiple equally spaced stations and analyzing much in the fashion of a time-series analysis.

Except where vertical segregation of the treatment (pollutant) in the water column is suspected, a vertical series of horizontally collected samples is not required and a deepwater system should be replaced by oblique tows. Vertical tows should not be considered except in exceptional cases.

Depending on the questions to be resolved, bioassay techniques using either "in situ" or laboratory simulated conditions should be considered as possible alternatives to near-field community studies.

Generally zooplankton collection devices can be categorized into three basic groups; water bottles, pumps and nets.

Table 1: Size of Common Zooplankton

Species	Freshwater		Marine	
	Habitat	*Size Range*	*Habitat*	*Size Range*
Protozoans	Few fresh	6-1000 micron⁴		−2.0 mm⁴
Ciliophora	Few fresh	22-600 micron		
Coelentera	Few fresh (Hydras)	< 20 mm		
Ctenophora				10-12 per cent
Platyhelminthes (flatworms)		1-30 mm	Inland waters	0.5–40 mm⁴
Nemertea (Proboscus worm)		< 20 mm	Inland waters	5 mm–6.5m⁴
Nematoda (Round worm)		<2.3 mm	Great variation	7 mm⁴
Nematomorpha (horsehair worms)	Pools	10-70 cm³		
Gastrotricha	Most fresh (shallow)	70-615 micron		
Rotifera	90 per cent fresh	80-1500 micron		
Bryozoa (moss animals)	Some fresh		Most species encrusting	
Chaetognatha (arrowworms)	Statoblasts	0.4–>1 mm	Statoblasts	
Annelida (segmented worms)			High salinity	Up to 40 mm
Oligochaeta	Most fresh	0.5–5 mm		
Polychaeta	Few fresh		Very few	
Hirundinea	Most fresh (standing water)	Adults 5 mm– 18 inches	Few marine	
Arthopoda				
Crustacea				
Branchiopoda	Most fresh	3 mm–30 mm⁴	Some marine	10 mm⁴
Cladocerans	Most fresh			
Ostracoda	(lentic waters)	0.2-18.0 mm³	Few marine	
Copepoda			Most estuarine/ marine in top of	
Calanoida	Nauplius-< 4 mm		Bottom sediment	
Cyclopoida	Nauplius-< 3 mm			<0.5-1 mm
Harpacticoida	Nauplius-1 mm			<0.5-1 mm
Ectoparasites	Some fresh	5-25 mm	Some estuarine	3.5-25 mm
Cirripedia			Estuarine/marine nauplii	
Mysidacea	Few in cold lakes	8-30 mm³		

Contd...

Table 1–*Contd...*

Species	Freshwater		Marine	
	Habitat	Size Range	Habitat	Size Range
Amphipoda	Some fresh	5-25 mm^3		
Decapoda	Some assoc. with debris	15-200 mm[1]		5-30 mm[4]
Insecta (aquatic)	Most fresh			
Mollusca				
Gastropoda	(mussel glochidium)	Adults 2-70 mm^3		8-80 mm trochophore
Pelecypoda	Some fresh	Adults 2-250 mm^3	Most have free swimming larvae	
Echinodermata				
Fish eggs/larvae	Eggs: 0.75-3 mm	400-500 micron		
	Larvae: 1.5 mm	Up to 2, 3 diameter		

Source: 1: Ward and Whipple; 2: Wilson; 3: Pennak; 4: Blackiston'sManual of common Invert.

Water Bottle

These have been historically used for capturing the micro-zooplankton, those organisms which will pass through a 200 micron mesh net. This sampling technique would basically capture rotifers and juvenile copepods and cladocerans, Since this is not an adequate representation of the zooplankton community in most situations, the water bottle technique is generally not acceptable.

Pump System

Another major category of sampling zooplankton gear is the pumping system which is enjoying increased acceptance recently due to large part of the greater accuracy in measuring water volume and in determining more precisely the depth at which the sample is collected. Pumping systems also permit continuous sampling of relatively large volume of water for net plankton; therefore, greater numbers of samples can be collected more efficiently than is possible with the net samples. These systems have a serious disadvantage in the avoidance of certain large macroplankton and its uses limited to those depths reached by the hose, A pump system should be considered for collecting meso-plankton from moderately shallow systems (50 m).

Nets

The final category is net samplers, which have a wide variety of instruments which are available for many applications. Since these devices incorporate netting, a brief description of net material will be considered. The most common type of net available is the simple locking monofilament mesh. The mesh pores which is important when in determining the size necessary for the zooplankton community intended for study. Although there is a tendency to adopt the smallest mesh size possible, there is a liability with small mesh sizes in reducing the collecting efficiency

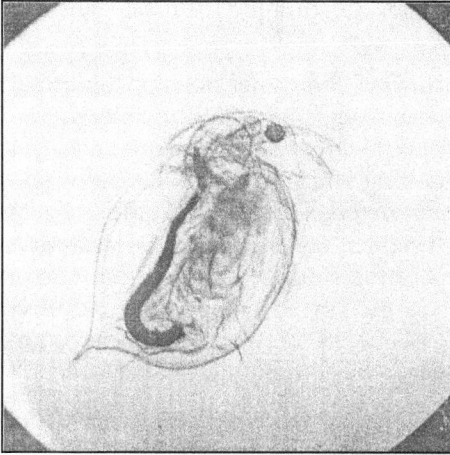

Daphnia magna with very little haemoglobin in its blood.

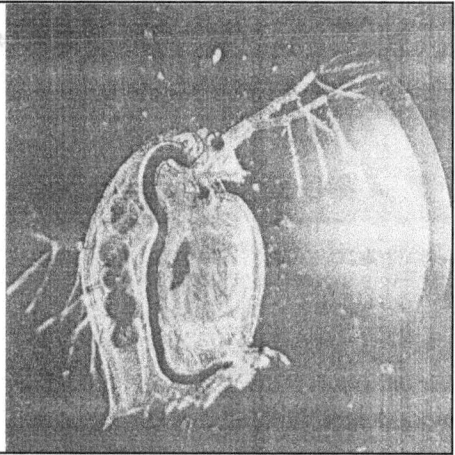

Daphnia magna carrying in its brood-pouch four eggs with haemoglobin in them.

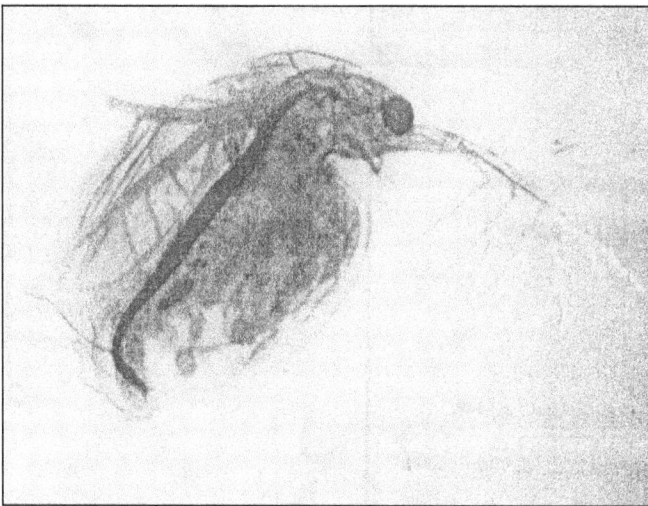

Daphnia magna, zooplankton of ponds with haemoglobin in its blood (x10).

of the gear as increasing equipment failure by the lower resistance to testing the finer mesh nets.In addition to the initial reduction in efficiency by the nature of the small mesh, the clogging rate of the gear is accelerated in that a 103 micron mesh net has been demonstrated to fall below 85 per cent efficiency within the first minute of towing. There is a rapid decline in water passing through the flow meter mounted in the net whereas the flow meter mounted outside the net remains relatively constant demonstrating the clogging effect apparent as a function of time. Table-1, indicates the types of organisms present in the zooplankton from various habitats and the size range of each which is necessary for determining the mesh necessary to implement a sampling program.

The size of sampling gear, and in particular, the dimensions of the mouth opening is directly related to the size of the organisms that will be unable to avoid capture. Therefore, a large mouth opening will yield larger forms in greater densities; however, it has the liability of decreased handling efficiency due to its bulky size. Therefore, in shallow areas, where small boats are necessary, small sampling gear, such as, a 0.1 m conical net or a water bottle becomes necessary and the limitations of the gear are then accepted. If the objects of the study were to concentrate on macroplankton then larger conical nets, such as, the 1 m net or one of the specialized mid-water trawl should be considered. The length of the net applied to these conical net frames is crucial and it is widely accepted that a mouth diameter to length ratio of 1:5 be used to increase filtration efficiency of the gear. Of the conical net

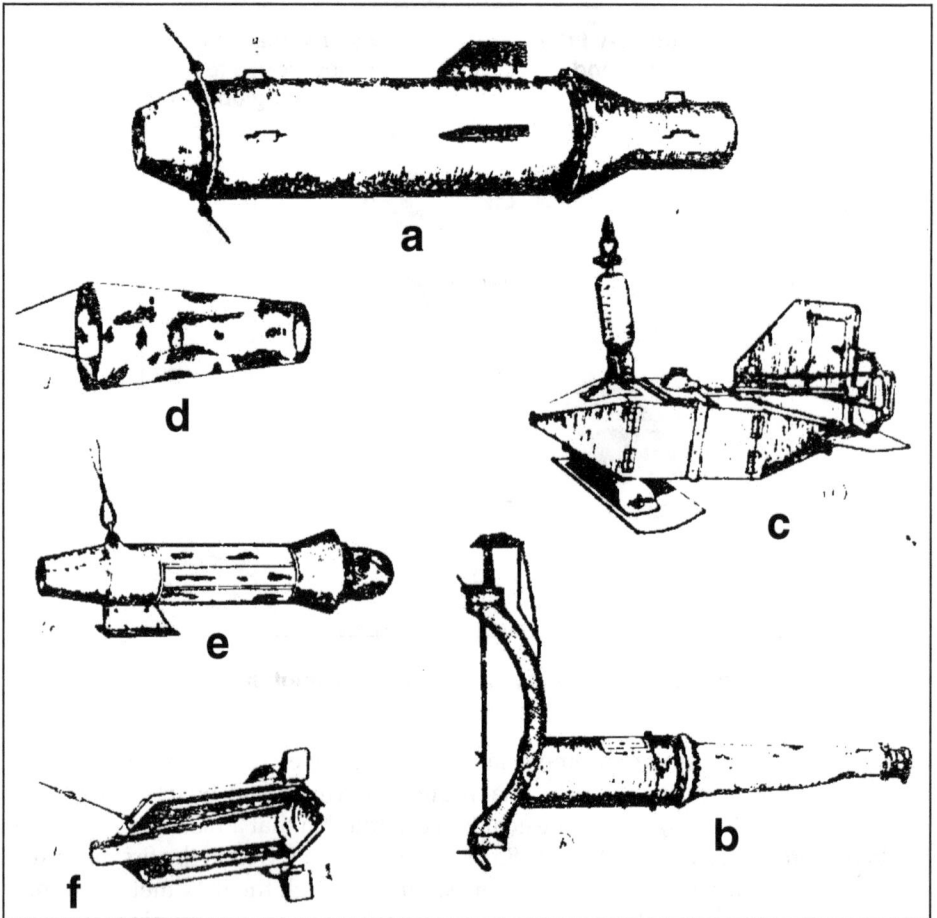

Selected Zooplankton Collect Devices.

a: Gulf III sampler; b: Clarke-Bumpus sampler; c: Hardy continuous plankton recorder;
d: Sheard high-speed sampler; e: Icelandic plankton sampler; f: Clarke jet net (section).

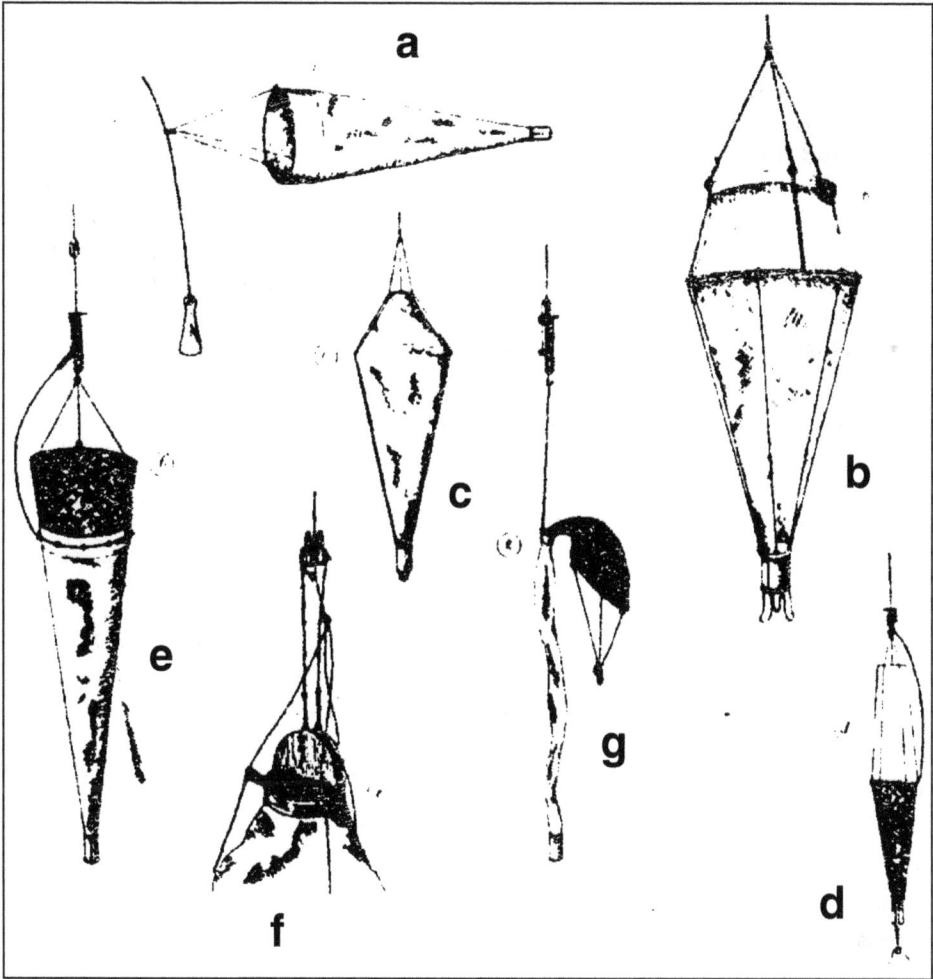

Selected Zooplankton Collection Devices.
a: Simple conieal tow-net; b: Hensen net; c: Apstein net; d: Juday net; e: Apstein net
with semi-circular closing lids; f: Nansen closing net open; g: Nansen net closed.

types, the bongo nets have proved particularly effective when compared with the Schindler trap and a pump system in the recent CEPEX study which were tested in a large contained water column. If the conical nets are to be used at any depth or in a multiple vertical series then closing nets are advised such as WP-2, in addition, it is also suggested that the string of the gear be stabilized with a depressor.

One of the most widely accepted zooplankton samplers which has been applied to freshwater, as well as estuarine and marine situations, is the Clarke-Bumpus Sampler which incorporates a closing device with a metered volume (flow water) register. It is a small, easily handled, and for most applications requiring integration of zooplankton patches, is a proven, reliable sampler. Other devices capturing small volumes of water are the Schindler and Juday traps, whose uses have been largely confined to lacustrine situations.

Most of the conical net samplers are designed to be towed at speeds of less than 1 knots. High speed samplers which are designed to be towed at speeds greater than 3 knots and up to 12 to 15 knots include the Miller high speed sampler, the Gulf Stream series sampler and the Hardy Continuous Recorder. These devices sample a very small horizontal section of the water column and are primarily designed to be towed over long distances with each deployment and therefore their use is restricted almost solely to pelagic marine environment or a large estuarine system. Finally, the last device to be considered is the Isaacs-Kidd mid-water trawl which is used exclusively in a marine pelagic situation when sampling large macroplankton.

Table 2: Zooplankton Collection Devices

Method	Marine			Estuarine			Riverine			Lacustrine	
Subsystem	A	B	C	A	B	C	D	E	F	G	H
Conical Nets											
WP-2 (0.94)		X	X	X	X	X	X			X	X
WP-3 (0.98)			X		X	X	X				X
0.10 m	X			X				X		X	
0.25 m		X	X	X	X	X	X			X	X
0.50 m	X	X	X	X	X						
1.00 m	X	X	X	X							
Closing	X	X	X	X							
Truncated	X	X									
Bongo (0.96)	X	X	X	X							
Tucker	X	X	X	X							
Schindler	X	X	X	X	X	X	X				
Juday	X	X	X	X	X	X	X				
Clarke Bumpus 0.88	X	X	X	X	X	X	X	X			
Water bottle	X	X	X	X	X	X	X				
Pumps	X	X	X	X	X	X	X	X	X		
High speed											
Miller	X	X	X	X							
Gulf Stream III	X	X									
Hardy continuous recorder	X	X									
Isaacs-Kidd Midwater trawl	X										

A: Intertidal; B: Subtidal; C: Pelagic; D: Lentic; E: Lotic; F: Tidal; G: Littoral; H: Profundal.

( ) denotes efficiency.

In order to collect quantitative samples, particularly with the conical net gear, a device to measure the flow of water through the net is necessary. There are several devices available ranging from the Japanese TSK devices to the more inexpensive though dependable devices avail from General Oceanics. These devices have to be calibrated periodically, as they tend to record smaller volumes of water as they age. Proper placement of the flow meter within the conical net is crucial. The greater velocity is recorded at the outer perimeter of the net, and the lowest velocity is recorded in the middle of the net with an increasing velocity gradient recorded between the center and the outside of the perimeter. In order to obtain an average velocity of water within the net, the meter is placed half way between the center and the outside perimeter of the net. It is also suggested that an outside flow meter be mounted in order to check for filtration efficiency which may deteriorate over a series of tows.

Benthic Macro-invertebrates

Benthic macroinvertebrates are animals, inhabiting substratum of lakes, streams, estuaries, and marine environments.Although very young individuals of many macroinvertebrates are quite small, by definition macroinvertebrates are visible to the unaided eye and are retained on a U.S. Standard No. 30 sieve (0.595 mm openings). Included among among the macroinvertebrates are aquatic insects, macrocrustaceans, mollusks, annelids, worms, corals, sponges, and numerous other aquatic and marine invertebrates.

The diversity and density (number of individuals per unit area) of macroinvertebrate communities are reasonably stable from year to year. However, seasonal fluctuations associated with the life cycle dynamics of individual species may result in extreme variation at specific sites within any calendar year. Most habitats with acceptable water quality and substrate conditions support diverse macroinvertebrate communities in which there is a reasonably balanced distribution of species among the total number of individuals present. Such communities respond to changing habitat quality by adjustments in community structure.

Macroinvertebrate community responses to environmental perturbation are useful in assessing the impact of energy extraction and processing wastes on natural water bodies.

Assessing the impact of a pollutional source generally involves comparing the macroinvertebrate community of sites presumed to be influenced by the source with that of adjacent unaffected sites. The procedure includes sampling and analyzing both communities and subsequently determining whether the presumed affected community differs from the non affected community. The basic information required for most community structure analysis is a count of individuals per species in the community. From the count data, the communities can be characterized and compared according to composition, density, biomass, diversity, or other types of analysis.

Station Selection

Survey objectives and the information desired should be clearly defined prior to selection of a sampling methodology. Selection of a methodology will ultimately depend on type of habitat to be sampled, that is stream, lake, estuary, or other habitat, and the type of pollution source to be evaluated, that is point or nonpoint source, discharge of wastes.

The following two criteria should be followed for selection of sampling stations at the very least.

1. Be sure that all sampling stations are ecologically similar. For example, the stations should be similar with respect to bottom substrate (sand, gravel, rock, or mud), depth, presence of riffles and pools, stream width, flow velocity, and bank cover.

2. Collect samples for physical and chemical analysis close to biological sampling stations to assure correlation of findings.

3. For a long term biological monitoring program, it is preferable macroinvertebrates be be collected at each station at least once during each of the annual seasons.

In general, the most critical period for macroinvertebrates in stream is during periods of high temperature and low flow. Therefore, if available time and funds limit the sampling frequency, then at least one survey during this time will produce useful information, Time is very important factor in bio-monitoring, especially in estuarine ecosystems.

Quantitative Sampling Devices

Grab Samplers

The most common quantitative sampling devices used in freshwater are Ponar, Petersen, and Ekman grabs and the Surber or square-foot stream bottom sampler, which are all described below. In addition, the orange peel grab, the Smith-McIntyre grab, and Shipek grab, useful in estuarine and marine habitats are described. Use of artificial substrate samplers in bio-monitoring and special procedures used for collecting freshwater mollusks are also described.

Ponar Grab

This device is widely used in medium to deep rivers, lakes and reservoirs. It is similar to the Petersen grab but has side plates and a screen on the top of the sample compartment to prevent loss of sample during closure. With one set of weights, the sampler weighs 20 kg. Word *et al.* (1976) reports that the large amount of surface disturbance associated with Ponar grab can be greatly reduced by simply installing hinges rather than fixed screen tops, which reduces the pressure wave associated with the sampler's descent. This sampler is best for sand, gravel, or small rocks with mud admixture and can be used in all substrates except bedrock.

Petersen Grab

This grab is used widely for sampling hard substrates, such as, sand, gravel, marl, and clay in swift currents and deep water. It is iron and weighs approximately 13.7 kg, but may weigh as much as 31.8 kg, when auxiliary weights are bolted to its sides. Primary purposes of the extra weights are to make the grab stable in swift currents and to give additional cutting force in fibrous or firm bottom materials.

The Petersen grab is made in several sizes from 0.06 to 0.09 meter square (0.6 to 1.0 square feet). These samplers should be modified by addition of end plates and by cutting large strips out at the top of each side and adding hinged 30 mesh screens as in the Ponar grab.

Orange-Peel Grab

This grab is designed for use in sandy substrates. It is a multi-jawed, round grab with a canvas closure at the top serving as a portion of the sample compartment. The 1600 square centimeter (100 square inch) in size is generally used, although larger sizes are available. The area sampled and the volume of material collected depend on the depth of penetration. This grab is suited to marine waters and deep lakes, where it has advantage over other grabs when sandy substrates are sampled.

Smith-McIntyre Grab

This heavy device is made of steel and its jaws are closed by strong coil springs. Chief advantages are its stability and easier control in rough water. It is bulky, requiring operation from a large boat equipped with a lifting device. The 45.4 kg grab can sample an area of 0.20 square meter (2.15 square feet).

Shipek Grab

This device is designed to take a sample of 0.04 square meter (approximately 8 inch by 8 inch) in surface area and approximately 10 cm (4 inch) deep at the center. It is usually used by geologists to collect small samples rather than by biologists. However it can be used in marine waters and large inland bodies of water.

Ekman Grab

This grab is designed for sampling in silt, muck, and sludge in quiet water with little current. The grab is made of 12 to 20 gauge brass or stainless steel and weighs approximately 3.2 kg. The boxlike part holding the sample has spring-operated jaws on the bottom that must be manually set (exercise caution in handling the grab – it can cause injury if accidentally trapped). This sampler is made in several sizes; however, only a tall model should be used, either a 23 cm or a 30.5 cm model.

Surber or Square-foot Sampler

Surber (1934) devised a useful device for macro-invertebrates sampling for wade-able, shallow streams. The square-foot frame is placed in the bottom and rocks, sand and other materials disturbed to a depth of two inches. The stream current carries the organisms and detritus back in to a 3 feet long net. Care must be taken to consistently dislodge bottom materials to the same depth on each sample for minimizing variability among replicates.

Artificial Substrate Samplers

Artificial substrate samplers are devices of standard composition and configuration that are placed in the water for a predetermined period of exposure and depth of placement for colonization by macro-invertebrate communities. Because many of the physical variables encountered in bottom sampling are minimized such as variable depth and light penetration, temperature differences and species substrate preferences, artificial substrate sampling effectively augments bottom substrate sampling. Like natural submerges substrates such as,logs, pilings, *etc.*,artificial substrates are colonized primarily by larval aquatic insects, crustaceans, coelenterates and bryozoans and to some extent by worms and mollusks. The organisms that colonize the artificial substrates are composed primarily of invertebrate "drift organisms" that are borne in the water currents, such as immature larvae and eggs. Because the colonization should be relatively equal in similar habitats, the numbers and kinds of organisms reflect the capacity of the water to support aquatic life.

The artificial substrates are usually positioned in the euphotic zone of good light penetration (one foot, 3 m) for maximum abundance and diversity of macro-invertebrates. The optimum abundance and diversity of macro-invertebrates for most waters in the U. S. occur at six weeks. For uniformity of depth, the samplers preferably are suspended from floats on 1/8 inch steel cable in not less than two replicates. If vandalism is a problem, surface floats or placement on the bottom may be required. Regardless of the installation technique, procedures should be uniform for a given study. For example, all samplers should be positioned for the same exposure to sunlight, current velocity and habitat type.

At shallow water stations (less than four feet deep), the samplers are installed so that the exposure occurs midway of the water column at low flow. If the samplers are installed in July when the water depth is about four feet and the August average low flow is two feet, the correct installation depth in July is one feet above the bottom.

Thus the samplers will receive sunlight at optimum depth (one foot) and will not be exposed to air anytime during the sampling period. Care should be exercised not to allow the samplers to touch bottom that may permit siltation, thereby increasing the sampling error. In shallow stream with sheet rock bottom, artificial substrates are secured to 3/8 inch (0.95 cm) steel rods that are driven into the substrate or secured to rods that are mounted on low, flat rectangular blocks. However, these must be securely anchored in the rock bottom to avoid loss during floods.

Before removal from the water, the sampler is enclosed in an oversized plastic bag (double wrapping) that is tightly sealed to prevent possible loss of organisms, or a large dip net is placed around (U.S. Standard No. 30 mesh) the sampler during removal. The samples may be disassembled and brushed in a pan of water in the field or preservative may be added to the bag consisting the intact sampler, which is later disassembled and brushed in the laboratory.

Although many different type of artificial substrate samplers have been tested, the Fullner modification of the Hester-Dendy multi-plate and the Basket sampler received wide spread use for water quality survey.

Multi-Plate Sampler

The modified Hester-Dendy multi-plate sampler is constructed of 0.125 inch (0.3 cm) tempered hard-board with 3 inch (7.5 cm) round plates and 1 inch (2.5 cm) round spacers that have center drilled 5/8 inch holes. The plates are separated by

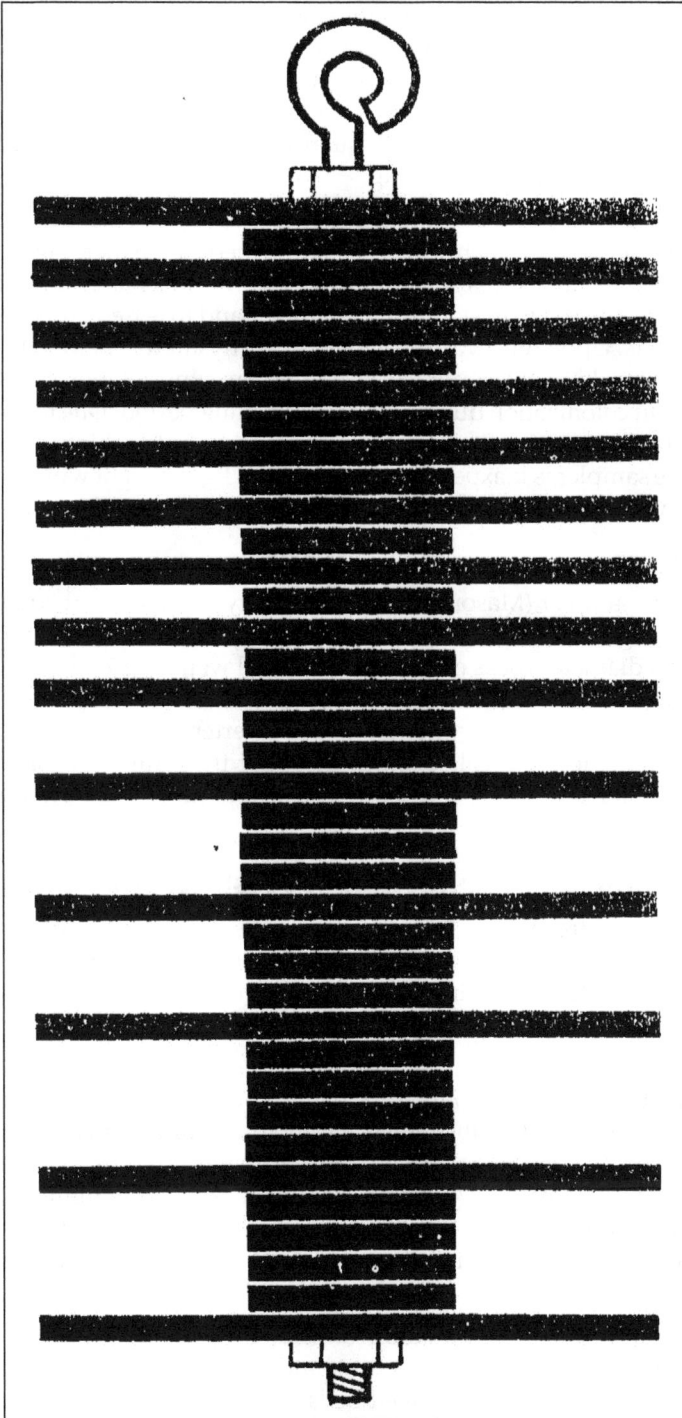

Modified Hester-Dendy Multiplate Sampler.

spacers on a 0.25 inch (0.63 cm) diameter eyebolt, held in place by a nut at the top and bottom. A total of 14 large plates and 24 spacers are used in each sampler. The top nine plates are each separated by a single spacers, and plate 13 and 14 are each separated by four spacers. The sampler is about 5.5 inch (14 cm) long, 3 inch (7.6 cm) diameter, exposed approximately1100 square cm of surface area for the attachment of organisms, and weigh about 0.45 kg.

When the samplers are suspended from the eyebolt, whether in strong currents or not, five pound weight, such as a brick, is attached by 0.6 meter wire to a ¼ inch turnbuckle. The turnbuckle is screwed tightly on to the shank of the multi-plate eyebolt. The weight serves to stabilize the sampler and to lessen undue disturbance to the organisms. Upon retrieval, the weight is gently cut free before the sampler is bagged. Care should be taken not to reuse samplers exposed to oils and chemicals that may inhibit colonization during the next sampling period. Due to its cylindrical configuration, the sampler fits a wide mouth container for shipping and storage purposes. The sampler is inexpensive, compact and light weight which are valuable attributes in water quality survey.

Basket Sampler

The basket sampler (Mason, et. al., 1973) is a cylindrical "barbeque" basket 11 inch (28 cm) long and 7 inch (17.8 cm) in diameter, filled with approximately thirty 2 inch (5.1 cm) diameter rocks or rock like material weighing 7.7 kg. A hinged side door allows access to the contents. The sampler provides an estimated 0.24 square meter of surface area for colonization by macro-invertebrates. The factors governing the proper installation and collection fit those for the multi-plate sampler. Some investigator prefer the basket technique because natural substrate materials are used for colonization.

Special Methods for Collection of Freshwater Mollusks

Importance of freshwater mussels and snails in environmental surveys are numerous. Once the site to be sampled has been identified, reference to be made to historical literature for determination of species that are likely to be encountered.

The best method for sampling large rivers is with a brail Six hundred meter long hauls (drag) should be accomplished where a single brail is used. Some fishermen use two brails; thus only three hauls would be required. Record the time for each haul. To be effective it should take about 20 minutes to make the hauls. If the hauls are too fast, the catch will be small. If a significant mussel population is found, then quantitative SCUBA samples should be taken. A minimum of 10 square meter samples should be taken at each station. All specimens should be identified to species, growth cessation rings counted, and measured for determination of population age structure.

Mussels found in small or medium streams that can be waded are often most numerous on bars where the pools break off into shoals. Often there are constrictions in streams at these points where weed beds can be found. Sample into the lower end of the pools, around the weed beds and in the riffles and fast-flowing waters. A Needham scraper or simply the hands are best samplers for these habitats. It is advisable to place a net below the area being sampled to catch small mussels that might otherwise not be collected.

Fish

Quantitative data on fish communities is highly desirable, however, it is often impossible to obtain, and is hindered by such diverse factors as inexperienced personnel to use the most efficient gear. Lack of adequate zoogeographic and ecological information, especially rare and endangered forms, indigeneous to hill streams. Recognizing the qualitativeness of fish collecting techniques, it seems imperative that an adequate base be established in all the states of the country. Since low pH and suspended solids appear to be the most damaging environmental factors to fish communities, their influence on behavior should be considered

The methodology for sampling and appraisal of fishery resources is a monumental task. A foremost factor to be considered is that the distributional status of fish fauna is lacking for many natural water resources (rivers, lakes, estuaries, reservoirs), and may be quite varied and complex.

Fish are mobile and difficult to collect. Gear is often highly selective and efficiency will vary with many factors, namely, life-stage, species, season and most importantly, the collector.

Attempts are made to summarize major definitive field techniques in the following chapter.

Chapter 3

Assessment of Fish Populations

Study Objectives

Environmental studies on fishes must be designed to accomplish specific objectives through the use of sound scientific principles. Studies should not be looked upon as merely academic exercises or programs to appease regulatory agencies or conservation groups. The staff employed to conduct such investigations must have adequate educational and professional experience both in field and laboratory procedures. Procedures followed need to be well documented. A bio-statistician should be consulted for the statistical design of the sampling program prior to its initiation. Taxonomic keys and regional references should be used for analyzing fish samples.

An outline of study objectives is a necessary and important beginning to any investigation. This is especially true for investigations of fish resources. The objectives should be realistic and clearly defined. The study itself should closely follow and relate to the objectives as outlined in preliminary planning. Once an appropriate plan is developed, this will have continuing influence on the investigation that follows. Certain tangibles such as sampling frequency, duration and whether or not the system is going to be continuously monitored, should be determined beforehand. For instance, if the investigation going to be one of a purely qualitative in nature, that is, determination of species present and their distribution, or is it going to serve as a long-term baseline for future monitoring? Objectives will determine how the survey is conducted and how the data is analyzed.

The specific objectives of a fisheries investigation may be related to either a pre- or post- operation of other activity. Some considerations depending upon the objectives of the study may be to determine the ecological, commercial or sport fisheries significance of species found in the study area; presence or absence and the ecological requirements of rare and endangered species, document species composition, relative numbers, seasonal distribution in the study area; document migratory patterns, feeding habits. External parasites and disease; or determine the spawning habits which includes the analysis of fish eggs and larvae. Behavior varies by species, age, season and time of day, and should be considered in life history investigations.

In fishery surveys, the number of stations and sampling frequency may vary considerably depending upon the habitat sampled. Riverine fish, in general, are much less migratory in nature; therefore, stations can be spaced at considerable distances. Oftentimes, depending upon the objectives of the investigation, there may not be a need to sample frequently. In lakes and estuaries, the migratory patterns of fishes must be decerned. Sampling localities are often dictated by accessibility.

Records

Of primary importance, precise records should be kept for each collection. The name of the collector, gear employed, exact locality, time involved in collecting, water temperature and date, are required as a minimum. Other recorded observations should include : width and depth of the stream; the presence and percentage of pool, riffle and run habitats, current, gradient and volume of water; plants observed, especially blue-green algae; and other properties, such as the presence of iron precipitate, coal washings or large bank of silt. Temperature has been considered the most important parameter influencing fish distribution. Vertical and horizontal profiles should be taken, especially in lake situations. This also applies to dissolved oxygen. pH should be recorded, especially in relation to potential acid mine problems

A reference or voucher collection of specimens should be maintained or deposited in a recognized museum. While important for the standardization of fish identification, a properly maintained voucher collection will be invaluable in legal proceedings. Under certain circumstances, it is advisable to take photographs;; all pertinent data relating to each frame should be recorded in a note book.

Preservation

Fish should be preserved immediately upon capture in 10 percent formalin. Large specimens should have their musculature and abdomen punctured. Specimens should remain in the preservative for two weeks, then be transferred to a 40 percent isopropyl alcohol or 70 percent ethanol solution after washing in water one time.

Gear

Appropriate gears should be used in the various habitat type. The quantitative aspects of appraising this important aquatic community have been largely ignored previously. Sampling and evaluation of fisheries resources is largely a qualitative task, especially in freshwater. In the marine environment fisheries scientists are not accustomed to this problem. Indeed, they may have years of data collected at the same locality by the same crew using the same trawl and can also rely on commercial fisheries reports.

In freshwater, available gear consists of seines, trawls, nets, electro-fishing units, ichthyocides and other miscellaneous techniques. The effectiveness of these major gear types are given below:

Seines

In acknowledging the limitations of a 5 X 10 seine, but found that efforts at any locality have collected the majority of species present. All available habitats should be thoroughly sampled and stop only after several seine hauls yield no new species is available. This obviously requires a familiarity of the ecological habits and taxonomy of fish. Bio-metritians may have difficulty living with this concept, but it should be recognized that while seines are standard field gear, they are basically a qualitative gear. Many records are based on juvenile specimens, thus, time of year is important in the collecting effort.

A common seine of 5 X 10 with1/8 inch mesh is recommended. It is the best all round gear available for sampling stream and shallow river areas. The 1/8 inch mesh is preferred over ¼ inch mesh since it more readily collects smaller species or life stages. A mesh size of ¼ inch is easier to use in high gradient streams.

Larger bag and beach seines have less utility in rivers than lacustrine environments, but can be used to various degrees of success depending on physical conditions and lack of obstacles. Since large fish are apt to outswim a beach or bag seine, monofilament gill nets can be used for seining if conditions permit. This is a highly selective technique, but may be of value for sampling large fish.

It is possible to approach the quantification of seine data under certain conditions. All other factors being equal, if the same experienced team is used for every collection, year to year, then quantifiable data is potentially possible. The experience of field personnel involved cannot be over emphasized.

Nets

All nets (gill, trammel, trap, hoop, *etc.*) are highly selective, and usually provide the best results when fished in combination. Mesh size, season, color of the net, size of the fish *etc.*, will influence efficiency. Nets should be fished at different depths to obtain information on vertical distribution. The use of nets is an acceptable technique for sampling lakes and low gradient rivers, although varying degree of success are obtained under river conditions and diurnal behavior cycles of fish. Net data are more readily quantifiable than seine data, and are usually expressed as catch per unit effort. Standardization of the gear, sample locality, sample period (time) and sample frequency will facilitate quantification.

Fyke and pound nets are essentially shallow water gear due to the difficulty in setting them. Fyke nets are used almost exclusively in some river fisheries and require a moderate current to be fished effectively. Pound and trap nets are used along shorelines to collect migratory species. Gill nets and trammel nets are used to take a great variety of fish, and can be adapted for use at different depths to obtain information on vertical distribution of fishes. Drift gill nets are sometimes used for pelagic species on large bodies of water. In the ocean, long streams of drift nets may be several miles long. Fish obtained by nets are often poor for use as diagnostic museum specimens, and those collected by trap nets are unsatisfactory for use in food habit analysis.

Electro-fishing

Electro-fishing has generally been considered as a rather effective technique and is often used in conducting population estimates. Cross and Stott (1975), in their experiments showed that the first of a series of replicates of electro-fishing through an area possibly causes a decrease in catchability of fishes on subsequent passes and effects may last for 24 hours and therefore, they suggested for re-evaluation of electro-fishing gear. Electro-fishing can be used in conjunction with seines in turbid water.

Icthyocides

Toxicants, conditions permitting, are undoubtedly the best tool for obtaining qualitative and quantitative samples of fishes. Relative abundance, diversity and biomass can be estimated more accurately than by other gears. If used by inexperienced personnel, control of toxicants and related problems may outweigh benefits received. The use of rotenone has been demonstrated in small streams, large rivers, and in impoundment surveys. Toxicants should be applied in conjunction to block nets and a neutralizer (namely, potassium permanganate oxidizes rotenone). The sample area should be surveyed for dead or moribund fish lying on the bottom or in eddy currents. Where possible, the use of SCUBA gear is advised.

Trawl

In most rivers and streams, trawling is impractical, if not impossible. However, trawl data collected in lakes, estuaries and oceans allow itself to be quantified (catch or biomass per unit effort). This is especially true if trawl data are available for a period of years from select locations. Trawl efficiency will vary with such factors as contour of the bottom, width of the mouth of the trawl, speed of tow, and depth of tow. Ellis and Pickering (1973) used a trawl equipped with electrical wings; however, they found it to be less efficient than a standard trawl.

Data Interpretation

The objective of the study will dictate methodology for evaluating data. Catch or biomass per unit effort is a rather easy concept to understand, but problems arise when such concept is employed as diversity or biota indices. Some individuals undertake data interpretation strictly as an exercise in mathematics rather than incorporation or considering biological principles. It should be recognized that diversity indices are a tool to assist in making judgement; however, they are not without fault. It is important to remember that "healthiness" of an ecosystem is a relative term dependent on the study area, season, sampling technique and objective assessment Diversity indices are based on the assumption that community structure is influenced and governed by environmental conditions, either natural or unnatural. In water quality assessment programs, one must distinguish natural community changes, namely, longitudinal distribution, from those caused by unnatural stresses, namely, pollution. Stations should be selected to minimize natural physicochemical variables and all available habitats sampled for their characteristic fauna.

Gear Recommended for Sampling Fishes by Habitat

	Pond	Lake		Stream	River		Estuary		Marine
		Shallow (< 50'0)	Deep (50 + ')		Shallow (< 20')	Deep (> 20')	Off Shore	Near Shore	
Seines									
5 X 10'		X	X	X	X	X	X	X	
5 X 25'		X	X		X			X	
150' Beach		X						X	
Purse			X				X		X
Nets									
Trap		X	X		X				
Fyke					X				
Gill		X	X		X	X	X		X
Trammel		X	X		X	X	X		X
Pound		X			X		X	X	
Trawl		X	X				X	X	X
Electro fishing									
Boat		X	X		X	X			X
Shore		X	X	X	X				
Backpack		X	X	X	X				
Ichthyocides		X	X	X	X			X	

As Wilhm and Dorris (1968) discussed for macro-invertebrate communities, a low index of diversity often occurs in stressed areas where a few species dominate. Conversely, unstressed communities are expected to contain more species, fewer individuals per species. This being reflected in a higher diversity index. Accordingly, this and other information indices are frequently used for comparing stable unstressed communities with those affected by adverse environmental conditions. All diversity indices have a common characteristic that numbers are generated for communities in study. Low values when compared to high values represent communities somehow deleteriously affected by the surrounding environment, either natural or man-made.

Data on occurrence and distribution of fishes; however, can be expected to be redundant in many instances. Fish are mobile and will migrate to and from stressed areas at will unless deleteriously affected. Individuals of a given species will congregate in a preferred habitat and juveniles of many species are especially are gregarious. Therefore, criteria established for one ecosystem will hardly apply to another. Numerical standards must be applied objectively.

The community concept is a true indicator of health. Factors other than pollution may affect the presence of a particular taxon, and complete life history and taxonomic relationships for many species have not been sufficiently researched to allow conclusions to be drawn with indicator species data. It is therefore reasoned that community composition offers a more reliable method of assessing stress conditions in aquatic ecosystems.

In closing, data analysis should include a blend of mathematical interpretations and sound biological principles. Techniques which may be employed include the Jaccard coefficient used by Hocutt *et al.* (1974) for analyzing qualitative fish data, ordination techniques, possible the Sequential Comparison Index, and the diversity indices of Wilhm and Dorris (1968).

Techniques of Fish Behavior Observations

Until recent years, biologists devoted considerable effort to the development of acute toxicity bioassays. The fact remains that LC 50 bioassays have limited biological significance. The biologists are not primarily interested in what kills organisms, but they are more interested in what sustains population levels and biological productivity. If LC 50 bioassays are used, Sprgue (1973) has updated the techniques through the use of computer techniques using probit analysis, which fit the probit line by iteration. Further development of this technique have been made with continuous flow apparatus and the care required in the bioassay.

Concomitant with the LC 50 type of study or slightly following the initiation of such study it became popular to study streams or lakes with "before and after" studies of the biota once the subject water had been exposed to pollutions. Rarely were these studies initiated before the exposure of the water to any given pollutants. Such studies are *post.hoc.* These are nothing more than correlations. They provide only an introduction to the problem, no firm ground on which to prosecute. What one is interested in is what does a pollutant do to the organism or a population of organisms at the time that it is active and as close to the environmental site and physiological site of effect as possible. It is proposed that this can be done and under some circumstances through study of behavior of the organism. It should be done at the time it is subject to the pollutant throughout the effect of that pollutant or completion of life history of the organism. Sublethal levels may interfere with a number of behaviors essential for population maintenance. These levels may prevent normal site or "territoriality" expressions, disrupt display, cause of abortion of eggs, depress dominance, *etc.* Under these circumstances, the populations may suffer from ecological death. Admittedly, such work is difficult and Stephan and Mount (1973) has presented some cause for doubt in use of this kind of bioassay. "Physiology, bio-chemical, histological and behavioral studies of fish have not achieved importance in pollution control work because experimental evidence of their biological significance is lacking. Any change from normal in a species is deleterious to the species can be extended to apply to whole aquatic ecosystems which would imply that any man-made change in the biological structure of an aquatic ecosystem should be considered detrimental – Changes from normal should not be considered detrimental to aquatic life until such changes can be shown to cause adverse effects on important organisms.

Behavioral Bioassays

The fact is that simple behavioral responses have been used and are being used. At Virginia Polytechnique Institute researchers have developed an automated biological monitoring system using opercular movement rate as an index of bluegill's response to chlorine and zinc sulphate. They have developed an automated system employing photoelectric cells and circuitry to determine the position of gold fish with respect to a field of different concentrations of copper ion. Robert Denoncourt *et al.* (1977) reported avoidance reactions at the population level of fishes exposed to cold, shock, and oxygen depletion. He found that carp and channel catfish would concentrate at the mouth of inlet streams and other refuges to escape the main river effect of depressed oxygen or sudden drop in temperature. Unfortunately, he was not able to continue the studies to determine what effect such crowding or exposure might have on survivality, growth, and reproduction. These last three requirements are the kind of data required to meet the objections to behavior bioassay. On the other hand, studies of mortality rates and growth rates leave much to be desired if one is studying the effects of pollution on fish populations. They are *post hoc*; therefore, only correlations.

Behavioral Response Bioassay

An attempt has been made to study normal behavior, particularly as it applies to those fish that seek cover. It has been demonstrated that smallmouth bass are intent seekers of cover. As complexity of the cover increases, that is, surfaces for thigmotactic responses, darkness of the cover and protection from current increase, the fish utilize the cover more extensively. Having established cover seeking as a worthwhile behavioral response in smallmouth bass subsequent trial using this response to determine how it is changed under decreasing oxygen levels and what effect lowering the pH has on the response. Smallmouth bass significantly reduced their use of cover and increased their activity when oxygen levels drop as little as 2 ppm and yet remain well above 5 ppm, the usual standard.

It appears that smallmouth bass do not change their use of cover nor increase their activity with a decrease in pH. However, it was shown that the fish do undergo changes in opercular rate, gapping and position under artificial cover when exposed to lower pH.

Obviously, if the fish does not respond to changes in pH, behavior cannot be used as a means of bio-assaying such toxicities, some of the pollutants have been shown not detectable by fish and therefore similar to the findings in pH, are PCB's and mercury, toxaphene, rotenone, arsenic, ammonia, hydrogen sulfide, phenols and creasotes.

Stephan and Mount (1973) feel that the ultimate criteria for chronic toxicity bioassays rests in measures of survival, growth, and reproduction. Even though they are attempting to relate field and laboratory work with studies of copper pollution, all measurements in the field are not necessarily cause and effect for their three criteria. Survival is primarily a behavioral based parameter. It is where the fish is, how well it is protected and to what extent it expends effort in the search for food. Growth may be more behavioral limited than food. Behavioral aspects of reproduction are basic to the numbers of fry that ultimately emerge. If the central nervous system is disturbed, such as has been reported with the exposure of fish to DDT, reproduction could be depressed at the behavioral rather than the population level. Ultimately, the measure is made at the population level but the cause can be primarily behavioral.

An attempt has been made to define with smallmouth bass and brown trout, the extent to which they use space and time. This can be done by watching mature individuals of a population in a stream section. Their diurnal activity as expressed in feeding, agonistic encounters, position changes and use of cover. It is hoped to relate that activity to growth differences among individuals and relate activity and site selection to such stimuli as cloud cover, food abundance, water velocity, turbidity, water temperature and season. Data are being collected on individual fish from 12 to 16 feet towers on the banks of the stream. With brown trout a technique of measuring have been developed for all these factors without ever having to handle, tag, mark. Or electro-fish a single individual. It is assumed that the results of this work will provide a basis as to what can be distinguished as "normal" behavior. Once this information is available, it may be possible to meet the objections of Stephan and Mount to go on for testing with pollutants to note significant changes in the behavior coupled with growth and changes in survival.

Bioassay Techniques

The ecological assessment of land use disturbances on aquatic systems has historically been performed by examination of the aquatic life communities in the impacted body of water. Effective assessments through such investigations frequently require detailed field studies involving sample collecting at several locations, time consuming laboratory analysis and repeat field excursions. Such extensive studies may be necessary to adequately document the environmental impact of the disturbance. The time requirement for the acquisition of this data may allow for irreparable damage to occur in the receiving waterway. Therefore there exists a need for a more rapid assessment technique that the resource manager can apply in attempting to evaluate potential impacts. One such tool is the modification of the research oriented toxicity test, commonly referred to as the "bioassay".

Sprague (1973) defined a bioassay as "a test in which the quantity or strength of material is determined by the reaction of a living organism to it". Bioassays have historically been used in research and experimental laboratories, usually involving "pure" or single compound testing, and have only recently been modified for field or remote testing use. Such application is becoming more and more in use, as documentation of environmental impact at the point of pollutant discharges offers considerable assessment advantages. An outline of the testing requirements and proper references necessary to successfully conduct such field bioassay investigations.

Toxicity tests or "bioassays" represent one of the best approaches available to field biologists for predicting the potential impacts of wastes on aquatic life.

Bioassay Tests

Toxicity testing falls into two basic categories; static and flow through tests. Static tests are usually short-term (acute), that is, lasting for 24 to 96 hours. Flow through tests may be either short-term or conducted as chronic tests lasting for several months. These tests have usually been conducted in fixed laboratory facilities far away from the pollutant source.

Adaptations of the static and flow through tests for their performance in mobile laboratories has been recently demonstrated successfully by the private industry and governmental regulatory agencies. Conducting these tests in mobile facilities can provide instantaneous data on the general toxicity of waste discharges on aquatic life. In essence, dilutions (log progression) of the waste are placed in test containers with test organisms. For example, the dilutions may be set up with full strength (100 per cent), a control (0 per cent) and a logarithmic series of concentrations in between these two levels. Common fish test organisms include the fathead minnow, bluegill, channel catfish, rainbow and brook trout. Commonly used invertebrates include *Daphnia* sp., *Gammarus* sp. (scud), mayfly nymphs, midge larvae, and mussels.

From an analysis of a short term toxicity test results, the biologist may able to predict the immediate effects of the discharge on representative aquatic organisms. Long term tests, more elaborate and complex in design, can provide a more reliable analysis to predict the chronic effects of a discharge on the aquatic life, for all life history stages. The conductance of these tests require strict control of variables, such as, flow, temperature, and water chemistry to clearly define the results of the discharge impact to the test organisms. Because of the complexity cost and man power requirements of chronic tests, this technique is not recommended for routine use by field biologists. To obtain reliable bioassay test results for fixed and mobile bioassays, the U.S. Environmental Protection Agency has developed specialized standards and testing protocols. These methods are described in details by Peltier (1978) and it is recommended that the investigator rely on those methods for standardization. The reference provides guidance to the specific requirements needed in facilities and equipment such as materials allowable for constructing containers, pumps, delivery systems, size of test chambers and rigid cleaning procedures. The kind, number, and source test organisms including disease treatment, holding and handling care, transportation, and acclimation are discussed in detail. The document also describes the required procedures for the type and collection of the dilution water and aging and treatments allowable. Specific requirements are also outlined for sampling the discharge and care of the test sample. The test procedure itself is described in the types of permissible tests and test conditions such as number of test organisms, loading of test organisms, test temperature, oxygen requirements and aeration, feeding and duration of the test are included. Treatment of the test results and reporting methods are also recommended.

Mobile Laboratory Bioassay

Gerbold (1973) describes many of the requirements and outlines both advantages and disadvantages of mobile laboratories. The operation of a mobile toxicity testing laboratory requires specialized training and familiarity with operation of equipment. In addition, the maintenance of the "flow through" needs to be monitored by at least two personnel, preferably a biologist and a technician. Involvement of the laboratory personnel during outfitting of the mobile laboratory is important to understanding of equipment problems and ultimately to the success of the test. If the test consists of a 96 hour test, then at least seven days are necessary from the on-site set up and acclimation of the test organisms through completion of the test and dismantling of the system.

A short fish "screening" test (24-48 hours) in a van type vehicles is useful in initial assessments of mining discharge impact. Careful attention should be given to standardized procedures. In the event the magnitude of impact should warrant additional documentation, then the more elaborate flow through test may be considered.

Bioassay Applications

Toxicity testing results and their application in understanding the impact of pollutant discharges have been demonstrated by many researchers. Bell (1971) subjected typical stream aquatic insects to various acid conditions and demonstrated the detrimental effects of low pH on the survival of this important fish food source. Cairns *et al.* (1971) experimented with subjecting a portion of a small mountain stream to acute acid stresses. The study demonstrated that benthic communities were reduced but retained no residual toxicity, thus allowing for re-colonization. A field study conducted by Reppert demonstrated how, by using fish cultures exposed to in stream conditions (impacted by mining discharges) toxicity tests could delineate the severity of acid mine drainage in an entire watershed. The resource manager can obtain rapid and realistic estimates of the severity of land use disturbances through properly conducted and controlled short time toxicity tests and by extrapolation through review of such past research predict the potential impact on the receiving aquatic resource. The short term test can be an effective tool in reducing the expansion of chronic conditions that have historically plagued the Appalachian Region.

Chapter 4

Habitat Evaluation Procedure for Aquatic Assessments

Methods for the Prediction of Impacts on Aquatic Communities

In the present era of energy shortages, oil, shale, coal mining, power and hydroelectric plants, irrigation and other resource-based developments continually place greater and greater demand on the Nation's water supplies, a resource already in critically short supply in many areas of the country. Concerned Federal and State agencies and conservation groups have been searching for appropriate methods that will provide quantitative and qualitative assessments of project development impacts in an attempt to protect diminishing aquatic natural resources.

In response to this need, Habitat Evaluation Procedures have been evolved for evaluating aquatic habitats.

Use of the Evaluation System

Aquatic habitat evaluation procedures have been developed to fulfill a number of needs in the area of resource development planning and environmental impact assessment. The objectives are to:

1. Develop objective methods to quantitatively assess base line habitat conditions for fisheries resources in non-monetary terms;

2. Provide an uniform system for predicting impacts on fisheries resources;

3. Display and compare the beneficial and adverse impacts of project alternatives on fisheries resources;

4. Provide a basis for recommending project modifications to compensate for or mitigate adverse effects on fisheries resources; and

5. Provide data to decision makers and the public from which sound resource decisions can be made.

Aquatic habitat evaluation procedures can be used at various stages in project planning and the data developed can play a significant role in the decision-making process. For most resource planning actions, the first level of activity concerns problem identification. The merit of any potential development is gauged largely by environmental and economic considerations. It is during this early stage of planning that aquatic habitat evaluation procedures can be very effective through quantification of baseline habitat conditions and predictions of the relative benefits (monetary and non-monetary) of all developmental measures, including those developed for the conservation of fisheries resources. Through early planning evaluation, the plan formulation process may be directed to produce an array of more environmentally acceptable alternatives than would be derived without these early planning insights.

Once alternative plans are formulated, aquatic habitat evaluation procedures can be used to quantify and display (or compare) their relative ecological impacts After the alternatives are assessed, recommendations can be made by the Service or other interested groups concerning the ecological acceptability or unacceptability of the various plans. From this assessment, several alternative courses of actions are available:

1. Concurrence with the project;
2. Concurrence with the project plan with certain project alterations specified and/or compensation measures incorporated into plans; or
3. Opposition to the project.

In summary, the aquatic habitat evaluation procedures will be used for;

1. Inventorying baseline habitat conditions;
2. Formulating alternative plans;
3. Evaluating alternative sites;
4. Evaluating alternative plans; and
5. Determining compensation requirements.

They will provide specific types of data at specified decision points along the planning process, including providing a means of identifying and assessing project alternatives early in the planning process. These procedures will not be the sole basis for recommendations on a given site, plan, or project. Rather they will provide ecological information for use in judging the trade-offs or merits of a proposal. Decision to oppose or recommend modifications for a given project plan are made on the basis of interpretations of the habitat evaluation analysis in connection with other information including resource abundance. State and local long-range resource plans, projected outdoor recreational needs, and other pertinent considerations.

Development of the Evaluation System

Aquatic habitats may be classified into aquatic eco-regions as a first step in the development of the aquatic evaluation system. It was postulated that the fish of the same species occupying different habitat types (that is, high gradient mountain streams as opposed to low gradient plains streams or of different climates) may be affected in slightly different ways by similar habitat parameters. The aquatic eco-region divisions are an attempt to define homogeneous response areas for widely distributed species of fishes. Eco-region boundaries are based on fish distribution, geomorphology, and climate information. Consideration of these data resulted in the preliminary aquatic eco-region map. Copies of the aquatic regionalization system have been sent to selected agencies for review and refinement.

The aquatic habitat evaluation system was developed as species oriented. Habitat evaluations will be based primarily on the ability of a stream reach or body of water to produce selected species of fishes. The evaluation system has also been designed to provide biomass estimates at reasonable confidence levels (70 per cent or higher) and intervals (plus minus 25 per cent) for those fish species where sufficient habitat requirements information is available. This segment of the aquatic habitat evaluation system is based on a trout biomass prediction model developed for Wyoming.

The evaluation system will utilize habitat parameters significant in determining the distribution and abundance of fish within each eco-region. These parameters will be selected, scored and weighted by species and life stage. Scoring of pertinent habitat parameters will be accomplished by use of response curves constructed by life stage for each selected species. Information used for the construction of response curves was obtained from the technical literature and the files of natural resource agencies. Response curves that determine parameter scores will be based on the effect of the selected habitat parameter on particular life stages of selected fish species. Optimal habitat conditions receive a score of 1.0 whereas sub-optimal conditions score decimal values of less than 1.0 down to zero when the tolerance limits are exceeded for the species. Response curves are constructed for selected species by eco-region.

In practice, selected habitat parameters of a stream reach or body of water being evaluated are measured and compared with the habitat requirements as indicated by the response curves of selected fish species. The software subroutine in the aquatic habitat evaluation model will score each of the pertinent habitat parameters to produce Habitat Suitability Index (HIS) scores by life stage for each species of fish evaluated. The suitability of the habitat has been evaluated, as indicated by HIS score., can then be examined by life stage, or a series of species HIS scores can be combined into a fish community HIS score. Multiplying the HIS scores by the total surface acres of the project impact area will yield a Habitat Unit (HU) index score.

The aquatic habitat evaluation model will have three subroutine levels. Level-1 will yield HIS and HU values as described earlier. Second level computations will permit the consideration of additional bio-ecological factors such as species of high interest or habitat uniqueness (namely, highly exploitable areas, wilderness streams and critical habitat areas) in producing adjusted HU values. The third level output from the model is an attempt to equate HU values among different habitat types and is used in evaluating compensation trade-offs between habitat types. Habitat losses may be compensated in kind, that is, replacement of specific habitat types Hu for HU, or it may be out-of-kind, that is, replacement of HU's of one type with HU's of an agreed-upon different type. Third level computations allow such factors as regional public preferences for particular habitat types or fish species, as well as various economic values, to be used as pertinent inputs into the model to derive regionally equivalent HU's for out-of-kind compensation purposes.

The aquatic habitat evaluation procedures outlined above should enable to examine the details of project impacts on the appropriate life history stages of fish species of interest. This will enable to more effectively protect aquatic habitats and to compensate for resource losses on an objective, quantifiable, biological basis.

Statistical Approach to Biological Predictions

The existing level of mathematical sophistication is clearly adequate to simulate complex ecosystem dynamics. Unfortunately, such simulations are only practical in those instances when the biological level of organization being mimicked is very strongly linked to obvious driving forces. One example of successful simulations of this sort is the modeling of eutrophication in response to nutrient loading. Here, of course, principles of uptake kinetics dispersion and advection, and of cell proliferation easily provide satisfactory system descriptions and predictions. Surprisingly however, because of the numerical nature of such exact models and the degree to which they rely on empirical facts, the models don't bear up well to generalization or transferability from one system to another, apparently similar one. As a further note of reservation, it may well be emphasized that the types of disturbances amenable to such simulation treatments can't be uniformly generalized either.

Alternatives to Simulation Modeling

Then what alternatives are available for the quantitative definition of ecosystem states and for the prediction of future states resulting from environmental change ? Quite clearly one practical way to gain tractability in the prediction of ecosystem response to disturbances (and to habitat restoration) is to give up the "cause-effect" point of view so mechanically represented by simulation modeling. At the same time, of course, another serious problem is alleviated; namely, that of having to know a very large number of details concerning ecosystem structure and function (rates and linkages).

What has been given up, and what is the alternative approach ? From the point of view of gaining new insight about biological laws governing the species interaction, very little has been lost by abandoning simulations dependent on empirical inputs and hypothesized links. One serious loss is that of a running accounting system keeping track of material and energy flow. How much of a loss is this from a predictive (but not from descriptive) point of view ? Very little loss of ability is incurred in prediction by the abandonment of Odum or Lotka-Volterra types of ecosystem models, since these models are all inadequate in defining any system completely enough from a "thermodynamic" point of view. Thus for the solution of practical impact prediction problem, such as the ones are confronting today, the viable alternative to the "cause-effect" approach is the more modest "correlative" one.

By admitting this posture of considerable ignorance as the practical guiding approach, statistical methods become the mainstays of impact prediction. Indeed as many of the other contributors point out, the gathering of even simple empirical data of sufficient accuracy and precision is a difficult enough and expensive problem. Thus the "correlative" focused statistical approach promises a better return on the effort than the acquisition of complex system behavior data to be used in simulations.

Statistical Approaches

It is quite clear that statistical approaches may treat biological and habitat properties individually, as well as together. Habitat alterations may initiate biological responses in certain components that amplify secondary (or consequential) effects to the extent that they exceed the response in the primary species affected. This is why parameter or input level changes in simulation model may fail to do justice to system response, unless secondary responses are built into the models *a priori*. Such fore knowledge is rarely the case, however.

Statistical prediction is a very specific concept to the statistician. By biologists and managers, the term should be understood in a more generalized sense. It may mean the following:

(a) Detection of an observed *effect* by hypothesis testing;
(b) Detection of ecosystem *changes* by reference classification to "known" or "undisturbed" or "bench mark" systems;
(c) Association of of habitat *changes* with biotic *changes*.

The techniques to be used for these perspectives include both univariate and multivariate statistics, depending on the complexity of observed "reality". In any case, multivariate analogues exist for all well-known univariate methods, and more. If possible, experimental approaches to ecosystem analysis should be multivariate to begin with, for complex systems being dealt with. In case the situation is simpler than expected, the multivariate schemes will quickly reduced to an univariate world. To elaborate a bit on the above analysis categories, multivariate analysis of variance and factor analysis are examples of hypothesis testing to be used for detection of effects. Classification methods include cluster analysis and discriminant analysis, with many variations on each of these. Multivariate analysis of covariance and extensions of discriminant analysis serve well in accomplishing the "correlative" goals of the statistical approach in ecosystem impact prediction. Furthermore, simplifications in prediction may also arise from multivariate analysis in the identification and singling out of indicator species associated with particular environmental (or disturbed) conditions. Thus with sufficient preliminary field experimentation and experience, considerable savings and increased accuracy may be gained by allowing designed multivariate analysis to select particular "predictor" components in the system.

Attributes of Biological Predictive Models

		Models on Effects		
Simulation (Mechanistic-Cause/Effect)			Statistical (Phenomenological-Correlative)	
Types	Application		Types	Applications
	* Physiological responses (Kinetics)		* Probit/Bioassay	
	* Population dynamics			* Species distribution
1 Deterministic	* Hydrodynamics		1 Hypothesis testing	* Habitat structure
2 Stochastic	* Biogeochemical (kinetics)		2 Classification/ Ordination	* Community structure
	* Bioenergetic			* Correlation with Environmental conditions
	* Species interactions with Water chemistry and physics			* Organism behavior
Data needs	Life cycle information on species (detailed in time and space) Lab/Field corroborative studies System definition and closure (System size) is critical		Standing crop information on species (comparatively simple/ empirical) Lab or field studies System definition is empirical/ more relaxed	
		Inferences on impacts		

Problem Definition

To effect survey programs designed to detect and predict impacts, a single important aspect of the problem must be recognized at the outset. This is the fact that the effects of all imposed disturbances are complicated by natural environmental gradients and fluctuations. Therefore, survey and controlled field experiment designs must be stratified to enable the separation of natural fluctuations from man-induced changes. It is also clear that the "correlative" approach does not require the comprehensive definitions used in "cause-effect" simulations of the system or subsystem to be examined. Nonetheless, the components included (biological and habitat) and the kinds of environmental effects to be tested or detected must be clearly defined, together with anticipated natural and other synergistic factors likely to interfere with the environmental diagnosis. All this points to the need for;

(a) The manager to identify specific problem at issue;

(b) The investigator within a particular discipline to address and focus on the problem defined in (a);

(c) The investigators to receive (or learn) the benefits of experimental design, which is the realm of the quantitative ecologist or bio-statistician;

(d) Maintaining the connections among managers, biologists and statisticians to achieve initial or modified goals as the problem evolves with ongoing analysis and updated experiments.

In spite of the best possible cooperation, only part of the problem is really solved so far. After all the statistical designs (not ex post facto manipulations), analysis and changes in experiments bear fruit, all that has been answered is whether a change or effect has been detected. The impact (or projected impact) is a subjective significance judgement to be made regarding theses changes —whether social, economic or political issues are being addressed.

From a purely scientific point of view, the "correlative" approach in field studies should not be sold cheaply as the avenue to the hackneyed "environmental impact assessment" edifice. In many ways, the statistical approach is superior to other quantitative techniques (such as simulation information measures and condition indices) as a way to discern environmental versus biological controls in the systems, and as a vehicle for recognizing generalizations in system behavior.

Chapter 5

Review of Fishes of the Region

The logical basis for work in marine and freshwater fishery biology, aquaculture, pathology, breeding, genetics and other lines of work with fish, is an exact knowledge of the kinds of fishes composing the fauna under investigation.

This course presupposes at least an elementary knowledge of ichthyology. If not done, one should become thoroughly familiar with the best available study of local fish fauna. Using the following list of the families and tables as a guide, and specimens and pictures the common and technical names of all the families of the fauna of the region can be learnt.

Class – Pisces

Sub-class

Order

Family

Genus

Species

To emphasize the importance of common names of fishes in dealing with lay conservationists, commercial fishermen, anglers and others and to help review the common fishes of economic significance of the region, the following table may be filled up with the aid of the instructor and the latest taxonomic references.

Review and Economic Classification of some Common Fishes of the Region

Scientific Name

Common name

Range and abundance

Game

Commercial

Fine

Coarse

Forage

Obno

Identification of Fishes and other Aquatic Organisms

Proper identification of fishes is of great importance for management, but it is often difficult to attain. Many fishes, such as, minnows are not easy to identify even with the aid of carefully prepared keys. Other fishes hybridize freely and the hybrids are not treated in most keys.

For identification of fishes there are several useful publications which together more or less the India, Mayanmer and Sri Lanka. It may be recognized, while working, that the faunas of certain regions (for example, North-East India) are less well-known than others. At least in the early stages of his work, an investigator should have his attempts at classification verified by someone qualified to do so. Sometimes the investigator may take his specimens to museums and compare them with materials previously classified by experts. At other times, the investigator may send series to such centers for identification, always with the understanding that only duplicates may be returned to him and that rare materials may be retained by the museum consulted.

Needless to say, scientific work requires the finest identifications. Only in this way may costly blunders be avoided, such as, the stocking of the wrong kinds of fishes, which has often been done. Furthermore, confusion in the literature of investigational reports may be held to a minimum and personal reputation maintained. No study in which organisms are mentioned by name attains its maximum value, be it a food study, a bottom fauna analysis *etc.* unless all forms are accorded their fullest possible identity.

Work Program and Record

As many as publications on the identifications may be examined and notes taken as to the content including reference to date, quality and research center at which the work was done. Such citation may be verified for accuracy. It may be noted that the list is composed of mostly of faunal studies and does not include revisionary ones. The latter very important for keeping nomenclature up-to-date, but are the particular subject matter of taxonomic ichthyology.

Development of a Life History Outline

The primary objective of current fisheries research is to develop management programs for increasing and sustaining a high yield of desirable species of aquatic animals. Basic to any such program is a knowledge of the life history and ecology of the organism primarily concerned. In order to acquient with the numerous features which may be studied in the lives of aquatic animals, a comprehensive outline for the investigation of the life history and ecology of the fish may be prepared. Base this outline on a study of available literature and asterisk those items which is considered to be of basic importance to the development of a management program. A complete and accurate bibliographic citation for each reference may be prepared.

Life History Outline

The details of outline may be arranged to the following headings. All points in it may not be applicable to any one species.

Description

Adults, Sexual differences, Relative size, Length, Weight.

Distribution

General, Seasonal, Migration, Arrival, Departure, Route of travel, Relative appearance of sexes

General Behavior

Character of water preferred, Character of bottom preferred, Manner of resting, Manner of swimming, Use of fins, respiration, Vitality, Association (in schools, *etc.*), Association with other fishes, Seasonal variations, Diuranal features, Nocturnal features.

Food and Feeding

Food habits, Feeding habits, Periodicity in feeding, Abstinence during spawning season, Abstinence during cold periods.

Reproduction

Age at maturity, Preliminary changes, Special male seasonal characters, Special female seasonal characters, Season of reproduction, Temperature of water, Preparation, Manner of sexual excitation, Nest-making, Parts assumed by sexes, Selection of place, Depth of water preferred, Manner of spawning, Time of day or night, Frequency of spawning, Behavior of males and females, Associated fish. Disposition of eggs, Number of eggs, Period of incubation, Retardation or acceleration of incubation by temperature, Care of eggs, care of fry, Period of care, Survival, Food of young.

Growth

Development, Successive changes, Size and age.

Pathology and Parasites

Competitors and predators

Economic Value

Population sizes, Value as food and otherwise, Manner of capture, Catch statistics, Conservation.

Legends

Beliefs and sayings connected with species.

Chapter 6

Fish Anatomy

A fishery biologist needs acquaintance with gross fish anatomy both for technical purposes (identifying and sexing fish and performing autopsies) and for public relations (answering the queries of fishermen).

The two most common types of fishes are the soft-rayed ones and the spiny rayed ones. The former are exemplified by such fish as carps, minnows which have only soft-rays in their fins. Perch, bass and others spiny-rayed and have one or more spines in their dorsal, anal and pelvic fins.

Anatomy of Soft-Rayed Fish (Trout – *Salmo iridius*)

External Characters

The body is elongated, compressed, thickest in the middle and tapering both to the head and tail. It is somewhat fusiform. The mouth is terminal. The upper jaw is supported by two freely movable bones, the premaxilla at the front and the maxilla behind. Both of these bones bear teeth.in a single row. When the mouth is opened a row of palatine teeth is seen internally and parallel to those of the maxilla; in the mandible of the roof of the mouth there vomerine teeth. The lower jaw or mandible is made up of a dentary bone on each side. The dentaries meet at a symphysis in front at the midline. Those symphysis becomes swollen and curved upward into a strong hook in breeding males.

The eyes have no eyelids; the flat cornea is transparent. A short distance in front of the eye is the double nostril which ends blindly. There is no external indication of the auditory organ.

On each side of the posterior region of the head is the operculum or gill cover with three bones; the opercular, subopercular and interopercular. The last is attached to the angle of the mandible. The preopercular is one of the cheek bones following the outline of the hyomandibular. The ventral portion of the operculum is produced into a thin membranous extension, the branchiostegal membrane supported by bones, the branchiostegal rays. The narrow area on the ventral surface of the throat which separates the two gill openings from one another is called the isthmus.

Inspiration is effected by the gill covers being moved outward about the middle of their length but sealed against the body posteriorly. During this process the mouth is open and water flows into it since it cannot enter from behind because of the closure of the operculum and branchiostegal membranes there. Expiration is brought about by the gill covers moving inwards. The mouth closes and oral valves seal the spaces about the teeth so that the water must exit over the gills and out the space between the operculum and body.

On the ventral surface of the body at the anterior base of the anal fin is the anus or vent.

The head extends from the snout to the posteriormost limit of the opercular membrane; trunk is from operculum to anus; the post anal region is the tail.

There are two dorsal fins. The anterior one is supported by soft articulated fin rays. The posterior one has no rays, it is a small, thick adipose fin. Other fins are the quite homo-cercal caudal, the anal, and the paired pectorals and pelvics; all of these are soft rayed.

The body is covered with small scales, but these are absent from the head and fins. A well-marked lateral line is present and is sensory in function.

Skin and Exoskeleton

The epidermis contains mucus glands and pigment cells. The scales are in pouches in the dermis. Each scale is a thin bony plate with surface markings. The scales are imbricated like the shingles on a roof. They are cycloid since the exposed margin of each is smooth. If they bore teeth, as in perch, they would be ctenoid.

Endoskeleton

The vertebral column supports the body axially and is made up of two kinds of vertebrae – trunk and tail or caudal. The first caudal vertebra has a hemal spine. The parts of a typical trunk vertebra are neural spine, neural arch, centrum, zygapophyses. The last named bear ribs in the trunk vertebrae but in the tail vertebrae they are presumably extended ventrally to fuse a hemal arch and hemal spine.

The posterior part of the caudal region is modified to support the tail fin. The last vertebrae have their centra deflected dorsally. Neural and hemal spines of the last several vertebrae are directed caudally, are broadened and flattened, lie close to one another, and support the caudal fin rays.

The skull is very complex and is composed of mingled bone and cartilage. Its parts are the cranium, hyoid apparatus, jaws and suspensorium, opercular apparatus, and branchial arches.

Anatomy of the Trout.

Dorsal and anal fins are supported by interneural and interhemal bones respectively. Pectoral fins are supported by cleithrum, supracleithrum, postemporal, scapula and coracoid. Pelvic fins float in the belly musculature on basipterygia.

The trunk muscles are arranged in zigzag myomeres. Special muscles of segmental origin operate the fins, jaws, eyes, and other movable parts.

The body cavity is partitioned into an abdominal cavity and a pericardial cavity.

Digestive Organs

The mouth leads into the pharynx from which the gill slits lead to the exterior and from which the esophagus leads to the stomach. A pyloric valve separates the stomach from the small intestine which continues caudally into the large intestine. Opening into the forepart of the small intestine are several blind tubes, pyloric caeca. Liver and spleen are present. The air bladder lies above the viscera and is open into the esophagus by way of the pneumatic duct; the condition is referred to as the physostomous one.

Respiratory Organs

Four pairs of gills, each with a double row of filaments, are present. The fifth gill arch bears no filaments. A hyoidean pseudobranch is present on each side anteriorly under the operculum. The respiratory surface of each filament is greatly enhanced by numerous lamellae.

Circulatory Organs

The heart consists of a sinus venosus, atrium, and ventricle. The ventricle leads into a bulbus arteriosus from which the blood passes into the gills by way of a branching ventral aorta. The blood is oxygenated in the gills and passes into the forward part of the body and into the caudal part; it goes anteriorly via the carotid arteries and caudally via the aorta. The blood supplies the various systems and returns to the sinus venosus. The red blood corpuscles are small, nucleated discs.

Nervous System

A more or less typical fish brain with ten pairs of cranial nerves arising from it is housed in the cranium. The olfactory sac is a blind cavity into which the nostrils open. The eye has a flat cornea and globular lens; the lens may be seen through the pupil (opening) in the iris. The ear has only middle and inner parts. Semi-circular canals are developed and otoliths are present.

Uro-genital Organs

The kidneys (mesonephric type) are of considerable size and run the length of the trunk dorsal to the swim bladder. Caudally the kidneys drain into a small urinary bladder which discharges into the urogenital sinus.

The testes of males are long. Smooth, whitish organs which extend throughout most of the length of the abdominal cavity. Each is continued posteriorly into a duct which opens into the urinogenital sinus. They lie just beneath the swim bladder.

The ovaries also run the length of the abdominal cavity and contain numerous ova. They have at least a granular appearance in contrast to the smoothness of testes. There are no oviducts in salmonoids but the eggs pass directly from the ovaries into the visceral cavity, thence into the genital pores which open into the uro-genital sinuses.

Anatomy of Sea Bass (Spiny-Rayed Fish)

Anatomical features as mentioned in the trout with the following differences;

1. Spines in dorsal, anal and pelvic fins. There are two dorsal fins; the first one with spines only, while the second one is with soft rays. In the anal and pelvic fins, spines followed by soft rays. The differences between spines and soft-rays are to be described, together with the number of spines and soft rays where they occur.

2. Oviduct- the relationship between the urinary bladder and to anus to be sketched.

3. Pneumatic duct is absent.

4. Presence of pharyngeal teeth in throat.

5. Number of tooth rows in jaw bones.

6. Exclusion of maxillary from gape of mouth and absence of teeth on maxillary.

Chapter 7

Fish Embryology

The study is designed to provide with a working knowledge of the developmental process in fish and to make acquaintance with terminology of life history stages to hatching. This information is useful in life history studies and in fish cultural work.

Sketches are required to be made of the followings:

1. Unfertilized eggs, stages of development with respect of maturity, size. Membranes to be identified carefully together with the development of ovarian egg for fertilization. This is important to ascertain brood fishes suitable for spawning.

2. Eggs just fertilized, swollen, changes after fertilization, formation of germinal disc and germinal ring, with the following stages of embryonic development, recording the development in hours/minutes after fertilization Diameter of the fertilized swollen eggs (sample of 25 eggs) are to be measured in mm along with the nature of swollen eggs, that is, pelagic/demersal/adhesive/non-adhesive.

   a. First cleavage
   b. Second cleavage
   c. Third cleavage
   d. Fourth cleavage
   e. Morula state
   f. Commencement of gastrulation
   g. Yolk invasion half completed
   h. Yolk plug stage

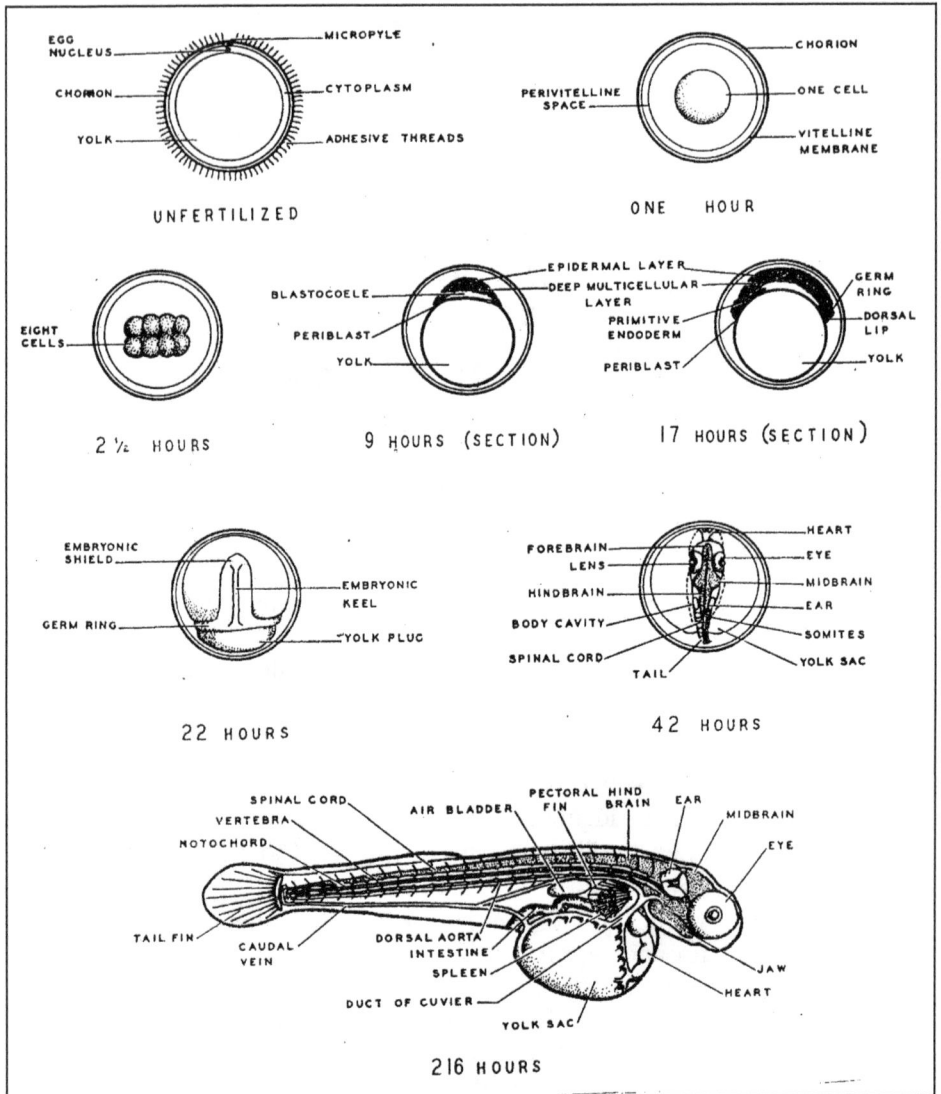

Embryology of the Mummic Hog.
(Redrawn from Solberg, 1938).

    i. Early embryonic streak stage

    j. Elongation of yolk mass commences, identify its cranial and caudal ends

    k. Appearance of myotomes

    l. Differentiation of head and tail regions.

    m. Appearance of optic cups and number of myotomes. Appearance of lens in optic cups

n. Elongation of tail from yolk mass and number of myotomes

o. Appearance of Kupfer's vesicle and number of myotomes.

p. Appearance of auditory vesicles and number of myotomes.

q. Appearance of auditory concretions and number of myotomes

r. Commencement of movement and number of myotomes

s. Indication of mouth

t. Pigmentation of eyes

u. Appearance of chromatophores

v. Appearance of pectoral fin buds

u. Hatching.

Besides the above study the volume of 1000 eggs is to be calculated in ml, calculating the number of eggs per ml. The weight of 1000 eggs is required to be determined. The color of the laid egg is also to be noted.

The water conditions, such as, temperature, pH, dissolved oxygen and carbon dioxide content is also required to be determined along with period of incubation.

Quantifying Fish Eggs, Fry and Young

In life history studies and in actual fish management it is often necessary to know the number of eggs, fry and young produced. The number of eggs must be known if survival is to be estimated- an important factor in fish production when related to fishing pressure and one that is inadequately investigated at present. Numbers of eggs produced are also of significance in fish cultural procedures since the size of brood stock, amount of rearing facilities, and extent of other equipment which must be on hand are dependent thereon.

The most accurate enumeration of fish eggs is by actual count, but this can be very tedious and time-consuming. When the actual counting of eggs is impracticable, approximate numbers may be obtained by several methods mentioned below. The results obtained by these methods may be recorded as:

Kind of Eggs (Species)

Average diameter of eggs in the sample ——————————————

Approximate number per quart (1 liter = 1.0567 quarts from von Bayer's table)

Method	Number of Eggs	Per cent of error
Volumetric		
Gravimetric		
von Bayer		
Actual count		

Volumetric Method

Count the eggs in a known volume (v), say 20 ml, which may be determined in a graduated cylinder. Then by measuring the volume (V) of the unknown lot, the total number may be calculated as follows:

x = total number desired, n = number found in volume (v);

Then, x : n = V : v

A somewhat different procedure may give more accurate results.

Count out 100 eggs and measure the volume of water displaced by them. A rapid way to count eggs is to have a piece of plastic with a known number of holes countersunk in it to accommodate the size of the egg in question. A graduate of small capacity is desirable for accuracy. Then measure the total displacement of the entire lot of eggs and calculate by proportion.

Gravimetric Method

Weigh a known number of eggs following removal of excess moisture. Then find the total weight of the lot and compute the total number by proportion.

von Bayer Method

Find the average diameter of the eggs by means of a small metal trough graduated in mm; remove excess moisture from between eggs. Make at least three determinations based on as many eggs as ruler in trough will permit and obtain average diameter. The von Bayer table giving the number per quart for eggs of various diameter may be consulted. These values may be changed to number per ml, by dividing number per quart, by 946.4. The number of ml in unknown lot may be found out by displacement and calculate number.

Canadian departments operating fish hatcheries have adopted standard methods of measurement for fish-culturists as follows (Rodd, 1947):

1. **Eggs** : Volumetric measurement in water based on three or more sample counts of the contents of a unit measure.

2. **Fry** : Computation by estimate checked against egg measurement with egg losses and visible fry losses deducted.

3. **Advanced fry** : In the early stages computation by estimate checked against egg measurement with egg losses and visible fry losses deducted. In the later stages, similar computation checked by sample counts.

4. **Fingerlings** : In the early and medium stages computation by weighing in water based on sample counts. In the later stages with larger fish, computation by weighing in water based on sample counts if the numbers are too large to count, otherwise by actual count.

   Computation by weighing facilities determination of the quantity and the cost of the food that goes into the production of a kg. of fish.

5. **Yearlings and older fish** : By count.

Sometimes it is necessary to make ovarian egg counts. To do this one must have entire ovaries, well preserved (in 20 per cent formalin to harden the eggs), and preferably taken just prior to spawning. If time and available materials permit, a mature ovary may be studied. Measure its total volume in cubic centimeters by water displacement and then obtain volumes of each of three or more well-spaced segments from it, each approximating about one cubic centimeter. Then separate out and count the mature eggs in each of the sample sections, segregating them from the other eggs and fibrous ovarian materials. Take the volumes of the latter remains and subtract from the volumes of the corresponding segments to get volume of mature eggs counted. Average the results for the three or more assays, compute number of fully developed eggs per ml, and then compute total number in the ovary.

Life History Stages after Hatching

There is need in fishery biology for the application of precise terminology to the early life history stages of fishes following hatching. In order to grasp the significance and distinctions of the following terms, study and sketch or photograph as complete a series for one species from newly hatched to adult and as many stages for other species may be made as material permits. A microscope may be used to obtain accurate results. In addition to making the sketches, label them fully and underline the label which identify the particular stage according to its proper life history term. On completion of the sketches, a list of characters for each species-stage, with distinguishable characters from other comparable ones should be made.

Larva

Developmental stages well differentiated from the later young and juvenile stages and intervening between the times of hatching and of transformation (loss of larval characters); the following two stages of larvae are commonly, but not universally, recognizable.

Pro-larva

Larva still bearing yolk, often called sac fry or fry by fish culturists.

Post-larva

Larva following the time of absorption of yolk but applied only when the structure and form continues to be strikingly different from the later stages. In salmonids and some other fishes the pro-larvae transform directly into young fish with the essential form and structure of adults and are called alevins or advanced fry by fish culturists. Advanced fry are ordinarily fish for a period of two weeks after complete absorption of the yolk sac.

Young

Usually taken to mean young-of-the year or a member of age group 0 and extending from hatching unit until January 1 in the Northern Hemisphere or July 1 in the Southern Hemisphere.

Juvenile

Fishes neither young nor adult but in the stages between. In a few fishes with very short life cycles the terms juvenile and young will be synonymous.

Adult

Mature fish, about to spawn, spawning, spent, or living to spawn again. In distinction, that stage of maturity or adulthood which is characterized by breeding condition of the gonads is called ripe; spent fish have just spawned.

Half-grown

A term to be loosely applied; because any precise distinction involves age determination, after which the following terminology of the life-history workers becomes applicable (Hubbs, 1944). In their practice the age-groups are numbered 0, I, II, *etc.* There is discrepancy as to time when group changes into the next one. Convenient is a division at January 1 in the Northern Hemisphere. Age-groups are not year classes; a year class is the group of fish spawned and hatched in any calendar year. For some species the term half-grown will apply to juveniles and for others to small adults.

Yearling

Member of age-group I, in second calendar year.

Two-year-old

Member of age-group II *etc.*

Observation of Larval Stages

The following observations may be extended at 12 to 24 hours intervals until the yolk is fully absorbed and the larva enters the post-larval stages.Differentiation during phases (hours after hatching) of larval life are to be made after 1 hour, 6 hours, 12 hours, 24 hours, 36 hours and 48 hours. Measurements of all characters are to be made in mm.

1. Total length
2. Height of body at pectoral level
3. Length of yolk sac
4. Maximum height of yolk sac
5. Number of pre-anal myotomes
6. Number of post-anal myotomes
7. Diameter of eye
8. Color of eye
9. Nature of auditory concretions
10. Position of mouth
11. Mouth functional or not

12. Barbels
13. Pectoral fin rudiment
14. Directive movement
15. Hind tip of notochord
16. Distribution of chromatophores
17. Commencement of feeding.

Post-larval Stages

Post-larval stages may be collected by rearing larvae under laboratory conditions and or in ponds. When reared under laboratory conditions care should be taken to feed the larvae with adequate live food so that their growth and differentiation would be almost normal. Samples from a pond would always be preferable to laboratory-reared samples.Post-larval features up to complete differentiation of all fins may be recorded as follows

Differentiation during phases (4, 7, 10, 15, 20, 25 days after hatching) of post-larval life are to be studied and measurements recorded in mm on the following characters.

1. Total length
2. Length of head
3. Diameter of eye
4. Height of body at pectoral level
5. Position of mouth
6. Nature of lips
7. Barbels
8. Hind tip of notochord
9. Embryonic fin fold
10. Dorsal fin
11. Rays in dorsal fin
12. Caudal fin
13. Rays in caudal fin
14. Anal fin
15. Rays in anal fin
16. Ventral fin
17. Rays in ventral fin
18. Pectoral fin
19. Rays in pectoral fin
20. Caudal spot
21. Lateral pigment line
22. Other characteristic pigment concentrations

23. Pigment on fins if any
24. Scales, appearance nature
25. Distinguishing coloration, if any, in live specimens
26. Any other feature

Assessment of the species for purposes of profitable cultivation would necessitate preliminary information on its growth and production when cultivated alone and also in combination with selected species of compatible habits. Such information could be gathered on the same line as indicated under the section on methods of cultivation.

Chapter 8

Fish Breeding

Reproduction in Fish

One of the basic factors of fish production is spawning success. Successful natural reproduction may sometimes be encouraged by environmental improvement, one of the tools of fish management. For this reason, the fishery biologist should know how to study the reproductive habits of fish and how to determine the spawning requirements of various species. Furthermore, such information is necessary when the stocking of fish in new waters is contemplated. Conditions unsuitable for reproduction have caused the failure of many plants of fish.

A fish species may be selected that spawns during a season and in a place where it may be observed. Many river fishes are particularly suitable for study where they occur in clear water. Lake and pond dwellers are also excellent for observation where the water is clear.

The notebook, camera, ruler, seine and other equipment may be carried to the spawning site.

Approach the spawning site cautiously. Determine the response of the fish to various degrees of movement and other disturbance on your part. Most fishes are not disturbed by motionless or quiet observers who become " a part of the landscape".

Determine the exact identity of the fish of primary observation and associated forms by (a) studying appearance in the water and (b) by careful seining to learn field characters. Sexual differences may be observed and record the description.

Observe the various elements of behavior of the fish in all aspects of reproduction. Repeated observations may be made of the same features. Keep copious notes. Be analytical. Record dates, times, location of activities, water depths, amount of current (if any), and related information. A safe rule is that nothing is too insignificant to take note of. This is insurance that when one begin to consider the continuity of the whole process, the separate elements will fall into their proper places and will be in true relation to one another.

Simple experiments may be designed; (a) verify as many natural observations as possible, (b) to disclose things that could not be learned by observation. A weir may be used to verify time of a spawning run and sizes, sex ratio, *etc.* of migrants. For example, one may feel confident from his observations that a male will accept only one female and one could verify this by introducing a second female experimentally in several trials. Or one may wonder if a female is recognized by color or shape. To answer this one might take male (different color) and round out his abdomen to look like that of a ripe female and offer him to the first nesting male. One could also experimentally offer spent females (same color as ripe females). In experimental design careful regard must be given to adequacy in number of trials and the elimination of variables.

Under no circumstances should it be assumed that activities of the sort observed, or other important ones, are limited to daylight hours without the study of the same fish during the same season at night. Some fishes spawn at night, others in the day, and still others both day and night; they may even shift from darkness to daylight or vice versa as the season progresses.

A sample of eggs may be gathered from the bottom or nest materials. From them percentage of fertile ones may be determined. Also some may be carried to the laboratory and reared through to obtain life history stages.

Photographs are particularly valuable in augmenting written descriptions of the reproductive habits of fishes. Good results have been obtained in daylight with camera out of the water by using an opposing screen to keep reflections of the sky from the water. Part of the flatness that characterizes daytime photographs may be overcome by flash photography at night, particularly if the flash bulb is placed in the water on a pole with an extension cord. Such lighting will bring out the contour of fishes, bottom, and associated structures.

Finally the observations may be compiled into a connected account of the reproductive habits of the fish studied. Compare your observations and conclusions with those of other workers on the same species and explain differences. This study will lead to the use of information of this sort for setting up the spawning requirements of a fish species for management purposes and will enlighten the additional information required.

Procurement of Stocking Material

Methods of procurement of stocking material may be considered on the basis of breeding habits of the fish species, namely, (a) species which normally breed in ponds; (b) species which could be induced to breed in ponds; (c) species which breed only in natural habitats, such as, lakes, rivers, estuaries, backwaters *etc.*

Species Normally Breeding in Ponds

Detailed records of the nature, extent and success of spawning have to be carefully recorded from critical study.

Genus and species _____

Size of breeding ponds : Length (m) _____Width (m) _____

Area (ha) _____Maximum depth of water (m) _____

Depth of water in the breeding area (m) _____

Area of marginal shallows (ha) _____

Nature of substratum in spawning area _____

Fish population in the pond at the time of spawning

Species	Nos.	Av. Length (cm)	Av. Weight (cm)	When Introduced

Water conditions and fish food during and after spawning

At Spawning Time	Days after Spawning			
	1 Day	3 Days	5 Days	10 days
Water temperature (°C)				
pH				
Dissolved oxygen				
Free carbon di oxide				
Total; alkalinity				
Plankton (Vol in 30 l water)				
Percentage of phytoplankton				

Predominant zooplankton

Extent of spawning

Breeding in the pond   Males   Females

Number

Average length (cm)

Average weight (g)

Total weight (kg)

Age (years)

Grown in the pond or introduced

If introduced, when _____ _____ _____

Potential spawning capacity of stock

Number of females_____

Average weight (g)_____

Average weight of ovaries (g)_____

Average total number of ova per female_____

Average number of eggs expected to be laid by a female at one spawning_____

Number of female breeders in the spawning congregation_____

Total number of eggs expected to be laid at the particular spawning_____

Total number of eggs laid as assessed by sampling_____

     In cases where eggs are scattered all over the sampling area, or where eggs are sticking on to weeds or some substratum sampling representative unit areas would be feasible. If the spawning area is also known, approximate number of eggs laid can easily be calculated from the average figure for unit area. In the case of nest-builders or mouth breeders such assessment is comparatively easy as selected samples can be examined but it may be difficult to return the samples for successful hatching.

Spawning Success

     Percentage of fertilization of eggs laid_____

     This is to be determined by sampling at fixed intervals after spawning until hatching.

Hour after Spawning	Sample Examined			Per cent of Fertilization	Remarks
	Total No.	No. of Developing Eggs	No. of Dead Eggs		
2-3					
12					
24					
36					
48					

Percentage of hatching _____

Approximate total number of hatchlings _____

Results of hatching under natural conditions may be compared with results under laboratory conditions where fluctuations may not be very marked. Samples of eggs taken from spawning ground are kept in shallow containers in the laboratory, with periodic change of water so as to maintain the dissolved oxygen level for successful hatching. Water conditions, such as, temperature, pH, dissolved oxygen, and carbon-di-oxide should be periodically determined. These samples would be examined at the same time as the field samples and the results compared.

The survival of hatchlings or larvae up to the post-larval stage or until the yolk is fully absorbed is separately estimated. In the case of carps, under natural conditions, the larvae will start moving almost within a day after hatching. Sampling under such conditions may not be representative and it would not be possible to assess the mortality during the period. Mortality due to factors other than predation may be estimated by keeping known number of hatchlings in small cloth or similar enclosures in the pond itself and examining these at desired intervals.

Enclosure Number	No. of Hatchling	Lapse of Time (hr) after Hatching	No. of Live Hatchlings	Percentage of Mortality
6				
12				
24				
36				
48				

In the case of nest builders, mouth breeders and those having other means of parental care, sampling selected post-larval broods would enable calculation of mortality/survival from the time of fertilization up to the post-larval phase. Predation is not a major factor in these cases.

When initial rearing is in the spawning pond itself, spawning success is assessed as the number of two-weeks old juvenile fish obtained per every million eggs laid. This is a total assessment by seining or dewatering the pond. When larvae are transferred to special nursery ponds this assessment would be more of rearing success than of spawning success.

Species that are Induced to Breed in Ponds

Common carp, though a pond breeding species, require specific inducement such as, egg collectors for successful breeding. It is well known that production of stocking material resulting from wild spawning of common carp is very limited and specialized techniques for breeding the fish have been developed for large scale production of stocking material. Detailed records of carp breeding and production of stocking material may be maintained as follows.

Particulars of breeding stock

Age (in years) : Male _____ Female _____

Pond/Habitat in which stock was reared _____

Male _____ Female _____

Sexes segregated : Yes/No

If "yes" for what period prior to release in spawning pond_____

Artificial feed, if any was given _____

Earlier spawning date _____

Present condition :   *Male*   *Female*

    Length (cm)

    Weight (g)

    Vent

    Abdomen

    Abdominal ridge

    Oozing

    General assessment

Breeding habitat

Pond/cistern/tub/netting happa _____

Size (m), length _____ width _____ depth _____

Source of water _____

Water Quality   Temperature   pH   D.O.   CO₂   Alkalinity   Color/Turbidity

Introducing Breeders

After spawning

Removal of eggs

For hatching

Depth of water (m) _____ Flow, if any _____

Brood stock

Time of day when introduced _____

Sex ratio by number _____ Male _____ Female _____

Sex ratio by weight _____ Male _____ Female _____

Weight (kg) of breeders	*Sex*	*Number*	*Total Weight*
Introduced	Males		
	Females		

Number of eggs expected to be laid by the females introduced _____

Egg collectors

Kakabans : Material used _____

Length (m) _____ Width (m) _____ Number of kakabans _____

Weeds Species _____ Quantity introduced (kg) _____

Spawning: Commencement (date/time) _____

End (date/time) _____

Spawned females

Number	*Weight Before Spawning*	*Weight After Spawning*	*Difference in Weight*	*Weight of Eggs Laid*	*Approx. Number of Eggs Laid*

Unspawned females

Number   Difference in weight, if any

The difference between the weight of the female before and after spawning will represent the weight of eggs laid, plus the weight faecal matter excreted during the period. While calculating the number of eggs laid an allowance for the weight of faecal matter has to be made on the basis of total weight of the breeders. Some idea of the weight of faecal matter excreted could be obtained from the difference in weight of the unspawned females.

Spawned females and males are transferred back to the stock pond. The unspawned ones, if in good condition may be kept, with fresh males, for breeding another day. If they still do not breed they are promptly transferred to stock ponds.

The eggs laid are to be sampled to determine the percentage of fertilization. Samples are to be measured as follows. Sample of at least 30 eggs are required to be measured.

Outer diameter (mm) of at least 30 fertilized eggs _____

Diameter (mm) excluding vitelline space of at least 30 fertilized eggs _____

Number of developing eggs_____

Hatching operations

Time of removal of eggs from spawning pond to hatching pond_____

Period (hour) between end of spawning and removal of eggs for hatching_____

Period (hour) between commencement of spawning and removal of eggs for hatching_____

Hatchery pond :

Length (m)_____Width (m) _____Depth (m) _____

Depth of water (m)_____

Flow of water (m/second) _____

Number of kakabans kept in the pond _____

Number of live eggs on them _____

Size of hatchery unit _____

Material with which made _____

Depth of water inside the hatchery unit (m) _____

Quantity of weeds distributed in the pond (kg) _____

Number of live eggs introduced in each _____

Hatching (date/time) _____

Period of incubation (hour) _____

Length of hatchlings in mm (sample of 30) _____

(1)   (2)   (3)   (4)   (5)   (6)   (7)   (8)   (9)   (10)   (11)   (12)   (13)

(14)   (15)   (16)   (17)   (18)   (19)   (20)   (21)   (22)   (23)   (24)   (25)

(26)   (27)   (28)   (29)   (30)

Yolk absorption completed by (date/time) _____ hour

After hatching.

Survival up to post-larval stage :

Hatchery unit _____

Number of eggs kept for hatching _____

Number of post-larvae obtained _____

Survival (per cent) from fertilization to yolk absorption _____

Length of post-larvae in mm (samples of 30) _____

Weight of 1000 post-larvae in gram _____

Rearing in ponds :

Preparation of the pond (fertilization, water filling, available natural fish feed_____

Pond number_____

Dimensions : Length _____ Breadth _____ Depth_____

Depth of water (m)_____ Flow of water (m/second)_____

Source of water _____

Details of fertilization, nutrient status of soil and water and fish food position to be recorded as detailed under habitat.

Insect control measures adopted (date/time)_____

Stocking : Date/time _____

Number stocked _____

Number per hectare of water area _____

Number per cubic meter of water stocked _____

Harvesting

In the case of ponds used as hatching-cum-nursery ponds, the number of fry obtained would represent the survival from the time of fertilization of eggs to the fry stage. In those cases where post-larvae were released for rearing the harvest would represent the survival during post-larval rearing only. In all cases representative samples are to be weighed and measured.

Date of harvesting _____

Rearing period (days) _____

Samples for Weighing (100 each) Sample-1   Sample-2   Sample-3   Average Weight

Per 100 in gram.

Measurements of total length in mm (samples of 50) _____

Production :

Pond number —————

Number harvested _____ Number per hectare _____

Survival percent _____

Total weight of harvest (kg) _____ Weight per hectare _____

Average daily per hectare _____

Calculated production/ha/year _____

Average length (mm) _____

Induced Breeding by Injection of Hormones

Pond reared Indian and Chinese carps (*Catla catla; Labeo rohita; Cirrhinus mrigala; Hypophthalmichthys molitrix; Ctenopharyngodon idella; Aristichthys mobilis; etc.*) are now induced to breed by the injection of pituitary hormones. The techniques have been standardized for field conditions and commercial production of fish seed. Proper recording of detailed observations on all aspects of this work will make more perfection of the standardized techniques.

Collection and storage of pituitary glands.

Pituitary glands are to be individually preserved and all particulars of the donor fish are to be recorded as below:

Donor Species	1	2	3	4	5
Habitat in which grown					
Caught on (date/time)					
Length (cm)					
Weight (g)					
Sex					
Stage of maturity					
General condition					
Date of collection of pituitary gland (PG)					
Interval between collection of fish and collection of PG (hours)					
How preserved during the period ?					

Donor Species	1	2	3	4	5
Size of PG in mm (length x width)					
Weight of PG (mg)					
Preserved in					
Preservative renewed on (date)					
Preserved at room temperature or lower temperature (°C)					
Used for injection (date)					
Period of preservation (days)					

Details of injections administered and results obtained may be maintained as below:

Recipient species _____

Habitat where breeders were matured _____

Recipient fishes	*Size*	*Sex*	*Length (cm)*	*Weight (kg)*
		Male		
		Female		

Breeding environment :

      Cistern/Pond/Reservoir/Canal/River _____

      Size _____

      Depth of water _____

Condition of water :

      Flowing/Stagnant/Fresh/Clear/Turbid/With bloom

At the time of	*Temperature*	*pH*	*D.O.*	*Carbon-dioxide*	*Alkalinity*
First injection					
Second injection					
Spawning					

Breeding enclosure :

Size _____

Material with which made _____

Depth of water inside _____

Pituitary glands used for injection :

Particulars	*For first injection*	*For second injection*
Donor species of glands		
Serial number of glands used		
Total weight of glands used		

Injection and spawning :

Injection	*First*		*Second*	
	Male	*Female*	*Male*	*Female*
Date, Time (hour)				
Dose (mg/kg)				
Weight of glands required (mg)				
Concentration of glands in extract (mg/ml)				
Weather conditions				

Interval between first and second injection _____ hours.

Date and time of spawning _____

Interval between second injection and spawning _____ hours

Eggs : Naturally fertilized/artificially fertilized _____

Weight of breeders :	*Female*	*Male*
Before spawning (g)		
After spawning (g)		
Difference in weight (g)		
Weight of gonads after spawning		
Whether spawning complete/partial		

Calculated number of eggs laid (weight of eggs laid x number of eggs per gram weight of ovaries).

Actual number laid _____

Vol. of Sample Cup	Number of Eggs per Cup			No. of Eggs per ml	Per cent of Fertilization	Total No. of Cups Measured	Total No. Laid
	Live	Dead	Total				

Diameter of eggs (inner/outer) in mm (sample of 30) _____

_____ _____

_____ _____

Total number of developing eggs _____

Percentage of fertilization may be checked by sampling as development progresses.

Species Spawning in Natural Habitat

Various species of Indian and Chinese carps, major species of brackish water food fishes like, Hilsa, Chanos and Mugil among others, come under this category. Location of seed concentration centers, timely exploitation of such concentrations, storage and transport to rearing centers are the major activities involved in fish culture.

Indian and Chinese carps breed in flooded rivers during monsoon months. Millions of eggs, hatchlings and early post-larvae washed down the current, are caught in specially devised nets, transported over long distance to nursery ponds and reared. Particulars of river spawn collection may be detailed as bellow;

River system _____

Name of collection center _____

Location _____

Condition in summer (dry/with water) _____

Substratum _____

Slope from bank _____

Width of collection ground _____

Date of first flooding during the season _____

Date of collection _____ time _____

Depth of water on collection ground _____

Flow of water _____ meter per second

Color of water _____

Water level _____ Rising/Falling _____

Number of nets operated _____

Size of nets used, length _____ width of mouth _____

Depth of mouth _____ Diameter at cod end _____

Length of tail cloth _____

Net fixed at what depth from surface _____

Net cleared at what intervals ? _____

Duration of heavy collection (hour) _____

Number of "hauls" per net during the period _____

Handling of spawn _____

Capacity of "cup" used for measuring spawn (ml) _____

Number of spawn per cup _____

Average number of cups of spawn obtained per net _____in _____ hour

Total collection _____ cups _____ number

Size of spawn collected (samples to be measured and recorded in tabular forms)

Collection of young ones of brackish water fishes :

Name of center _____

Date of collection _____

Species _____

Location of center _____

Within tidal influence : yes/no

State of tide at collection _____

Depth of water at collection ground _____

Method of collection _____

Size of material collected (by examples) _____

Quantity/Number collected _____

Duration of collection (hour) _____

Temporary storage : container _____

Volume of water _____

Number of fish per container _____

Duration of storage (hour) _____

Mortality during storage _____

Chapter 9

Methods of Cultivation

Stocking material are transported as spawn, larvae, fry or fingerlings. When the distance involved is short and the number required is small, transport is easy in open containers. When large quantities are to be transported over long distances, methods of economic transport ensuring maximum survival are to be adopted. Transport requirements of different species of fish differ from one another and also from stage to stage and this phase of activity therefore calls for early standardization of techniques so as to avoid wastage. Experiments and observations on transport of young fish may be recorded in detail on the lines indicated below :

Species transported _____

Nature of stocking material transported : eggs/larvae/post-larvae/fry/fingerlings

Size (length in mm of a sample of 30)

Range of length (mm) _____ Average length (mm) _____

Weight of 100 (g) _____ Average weight (g) _____

Approximate age (days) _____

Collected from _____

How collected _____

Date/time of collection _____

Conditioning, if any, (duration in hour) _____

Interval between collection and packing (hour) _____

	Temp.	Water Condition			Alkalinity
	°C	pH	D.O.	CO_2	

In the rearing pond

In the pond where

conditioning was done

Of water used for

Packing

Time of packing _____

Shape/size of container _____

Volume of water in container _____ l

Number of young fish introduced _____ weight _____ g

Volume of oxygen introduced _____ l

Under what pressure ? _____

Total weight of packed container with fish and water _____ kg

Mode of transport : by road/by rail/by air

Duration of transport : (Time taken from packing to opening of container for release in pond _____ hour

Date	Time	Water Condition in Container on Opening				
		Temperature	pH	D.O.	CO_2	Alkalinity

Condition of fish on opening container at destination _____

Mortality (No.) _____ Percentage _____

When short distance transport in open carriers is undertaken the additional information to be recorded is only the details of complete or partial removal of water en route. This should include the quality of water used also.

Besides, the basic preparation of the habitat for rearing, other aspects of cultivation will be supplementing natural food by artificial feeding; manipulation and preventative/curative measures against infection and disease.

The main principles in selecting artificial feed are easy availability and low cost. Whatever feed is selected its use should be on a common pattern so that the results would be comparable.

Phase of cultivation : Nursery pond/Rearing pond/Stocking pond/Brood or stock pond.

Artificial feed _____

Nutrient value (percentage)

Feed item Water Protein Fat Carbohydrates Total digestible nutrients

Size of pond _____

Water content _____ cubic meter _____

Water stagnant/flowing _____

If flowing, rate of flow _____

Stock of fish : Number _____ Average length (cm) _____

Average weight (g) _____

Total weight _____ kg

Daily ration of artificial feed : actual quantity _____ kg

As percentage of weight of stock _____

Mode of feeding _____

Frequency of feeding (how many times a day) _____

The effect of continued artificial feeding on water quality, plankton and fish has to be periodically determined. Weekly determination of water quality and plankton and fortnightly sampling of fish to determine the increment in weight due to feeding may be made. The quantity of daily feed to be given would then be determined every fortnight on the basis of the weight of stock at the end of each fortnight. After a specified period of feeding the results may be assessed as follows :

	1st Fortnight	*2nd Fortnight*	*3rd Fortnight*
Area of the pond			
Stock introduced (No)			
Average length of stock at commencement of feeding (cm)			
Average weight (g)			
Total weight of stock in the pond (kg)			

	1st Fortnight	*2nd Fortnight*	*3rd Fortnight*
Duration of feeding (days)			
Feed given per day (kg)			
Total feed given during the period (kg)			
Number of fish recovered at the end of feeding period			
Percentage of survival			
Average length attained (cm)			
Average weight attained (g)			
Average increase in weight (g)			
Total weight of stock (kg)			
Net increase in weight of stock (kg)			
Average increase in weight in the pond per day			
Average increase in weight per ha per day			
Calculated annual increase in weight per ha/year			
Conversion ratio (weight of feed : weight increase)			

Artificial feeding may be for assessing the value of particular feeds or for assessing the response of selected species to particular feeds or for increased production. Detailed records as indicated above would facilitate analysis of data from all required aspects.

Manipulation of stock in the pond is an important step in proper cultivation. The stock to be introduced into the pond has to be determined on the basis of the inherent fertility of the pond, the normal growth of the species stocked and the quantity of manure or artificial feed proposed to be introduced. Different combinations of species at different stages and in the different densities have to be tried before a suitable combination can be arrived at.

Pond No. _____ Area _____

Date of stocking _____

Species Stocked	*Number*	*Average Length*	*Average Weight*	*Total Weight*	*Stock per ha*	
					Number	*Weight*

Details of preparation, manuring and feeding, if any, are to be recorded.

Sampling would be done every month so that the growth achieved may be assessed. If the growth becomes poor it is an indication that the stock may have to be thinned out either of all the species or only of the particular species showing poor growth. Details of such fish removed and the effect of this removal on the left over stock have to be carefully assessed during subsequent samplings. At the end of the specified period of growth the entire stock is harvested and particulars noted for ascertaining the survival and growth of various species.

Particulars	*Species in the Combination*
Average length at stocking (cm)	
Average weight at stocking (g)	
Number stocked	
Total weight (kg)	
Period of growth (days)	
Date of thinning	
Number thinned out	
Average length (cm)	
Average weight (g)	
Total weight (kg)	
Harvesting date	
Total period of growth (days)	
Number harvested	
Average length (cm)	
Average weight (g)	
Total weight (kg)	
Total number recovered including those thinned out earlier	
Percentage of survival	
Average growth (weight)	
Per day per ha	
Per annum per ha	
Per annum per ha for all species together	

For a proper assessment of the growth recorded it is necessary to correlate the same with the manure applied and the artificial feed given during the period. The nutrient value of manures and feed have to be considered for this purpose. The performance of individual species has to be critically examined in order to decide about necessary variations in density or combination.

The stock of fish in the pond, at all stages is subject to hazards of predation, parasites and disease. Parasites and diseases are of diverse types and occur in all types of habitats and conditions. Detailed records of the environments, the state of the affected fish individually, as a species and as a constituent of the community, the symptoms of the disease or attack, the stages attacked or affected and measures taken have to be carefully maintained in order to attempt some sort of standardization in this aspect of work.

Date _____ Pond _____ Locality _____

Fish population in the pond (Tabular statement of species with size)

Other organisms present (Tabular statement)

Species of fish affected _____

(Tabular statement of affected species, size, number affected, *etc.*) _____

Nature of disease (Symptoms to be described in detail) _____

Disease-causing agent (Description) _____

Mortality of fish, if any _____

Water conditions in the pond (Tabular statement) _____

Curative/preventive measures adopted _____

Results of treatment _____

The chief principle and aim of fish culture is increased production of selected species of fish through judicious manipulation of the environment. While the response to various treatments and techniques in different habitats in different countries, by different species may be different, the methods employed and the results achieved in different localities should be comparable so as to avoid duplication and waste of effort and funds. Standardization of techniques is necessary not only for the better utilization of knowledge but also for quick progress in our knowledge of this important field of activity. The fundamental step in such an attempt is ensuring some uniformity in the manner in which basic data on various aspects of fish culture are recorded and maintained.

Chapter 10

Food Habits of Fish: Food Analysis

Fishes have evolved a wide variety of ways to obtain their food and meet their energy and nutritional requirements. As aquatic vertebrate, fish encounter potential dietary items that swim actively through the water, float microscopically at different depths, lie cryptically on the bottom, remain attached to the substratum, contain noxious or tough materials, concealed themselves in burrows and crevices, or exist as particles of uncertain organic origin. These items range enormously in size, structure, digestibility and nutritional content and, therefore, require that fishes as a group exhibit a broad array of different capture devices and digestive mechanisms if they are to exploit these foods successfully.

Collectively, fishes are opportunists in terms of food and feeding and show a high degree of variability both within and among species in capturing and processing food. Thus the type of fish feeding and digestion can be split finely and assigned to a broad array of categories. Conversely, in recognition of expansive and overlapping trophic diversity and differential digestibility among fishes, a less restrictive classification of fewer categories may be justified. In the latter sense, fishes can be classified according to the type of feeding mechanism used (biters, suction feeders), the kind of food they consume (herbivores, carnivores), their position in the food chain (primary consumer, secondary consumer) or even to the way in which food is digested (fishes with muscular stomachs or with hindgut fermentation chambers).

Capture and Ingestion of Food

Once they have detected and located prey, fishes use a wide range of behaviors and mechanical actions to catch and swallow their food. Fish feeding according to action of jaws and mouth, to body shape of fishes, to particular types of behavior, to the size of the prey, and to the types of fish consumed can be categorized.

Jaw and Mouth Action

Based on functional and ecomorphological analysis, feeding methods of fishes can be divided into three major types, biting, ram feeding and suction feeding. In biting the oral jaws are used to remove a portion of a larger organism or tear the attached prey from the substratum. The essential elements here are relatively robust jaws with cutting teeth, restricted jaw mobility and enlarged abductor muscles. Examples of food items obtained by biting include macro-algae attached to rocks, sedentary polychaetes, coral polyps, and clam siphons. If the food is especially large and tough, some fishes shift to spinning – in which the prey is grasped and the body rotated rapidly about the long axis to tear small pieces from a larger item. Predatory sharks and anguillid eels feeding on fish or other large prey and herbivorous pricklebacks (Stichaeidae) consuming pieces of sea weeds are known to resort to spinning.

In ram feeding, the fish opens its mouth and swim through a concentration of prey, whereas in suction feeding the fish expands its buccal cavity to create sub-ambient pressure, which draws a jet of water along with the prey. These two types of feeding may involve upper jaw protrusibility, a major feeding specialization seen among more highly derived teleost fishes. Specializations for ram feeding include a large mouth area, a large ratio of mouth area to buccal cavity volume and a body morphology suited to acceleration which involves a moderately elongate trunk, thick caudal peduncle and posteriorly placed dorsal and anal fins. Examples of food items most vulnerable to ram feeding are those prey that are elusive because they are adept at swimming and include fishes, shrimps, and calanoid copepods. Both ram and suction feeders are specialized for speed over force, having less robust but flexible jaws equipped with holding rather than cutting teeth and equal development of abductor and abductor jaw muscles.

Suction feeding is a feature of most teleost fishes and considered to be the most versatile type of feeding known among vertebrates. This technique is used in some form by particulate plankton feeders, chasers of fast moving prey, algal scrapers, invertebrate pickers, extractors of crevice-dwelling animals, and fin biters. Specialization for suction feeding include a small mouth area, a small ratio of mouth area to buccal cavity volume and a body morphology suited for maneuverability, which involves a laterally compressed, relatively deep body and laterally positioned pectoral fins for fine tuning the fish's position before the strike. Examples of food items most vulnerable to suction feeding are those prey that grasp the substratum, such as, limpets, isopods, crabs.

Food relationships function at least in part to determine population levels, rates of growth, and condition of fish. They serve as a partial basis for determining the status of various predatory or competing forms. For any species, they change with the seasons, with life history stage, and with the kinds of available food.

Since food studies show details of the ecological relationships between organisms, complete identifications of food items are requisite; this entails good working facilities. It is impossible for one person to accord this treatment to all of the remains that he finds; it is only with practice and by much comparison with reference collections that one may become proficient even for a few groups of organisms. The aid of other specialists must be used for identification in order to obtain the highest quality of food analysis.

Food studies based on contents of digestive tracts or of droppings merely show what a fish eat. Such facts when related to feeding habits and accompanied by ecological studies may constitute important management tools. A study of feeding habits may be of great help in reducing limitations imposed on food analysis by the differential rates of digestion of various food items. If it is first determined when a fish feeds, specimens may be collected close to or during that time and thus they will have been taken when they contain relatively whole, undigested materials, whether the organisms eaten are soft- or hard-bodied.

Because of differential rate of digestion, variation in size of food organisms, persistence of hard parts, and the usual inadequacy of samples, data from no one method of analysis may be solely relied upon.

Sample Record Form for Fish Food Analysis

Species _____ No. _____ Organ _____ Vol.
_____ ml. (1)

Sex _____ Age _____ Size _____ Condition
_____ (2)

Exact locality _____ (3)

Collector _____ No. _____ Date _____
Time _____

Food item	No.	L. found	L. alive	Vol.(ml)	Per cent Remarks
		(4)	(5)	(6)	

Explanatory Notes

1. Record total volume of food in the contents of the organ or organs studied.
2. Condition refers to relative fullness expressed to the nearest estimated tenth, for example, 0.6 full. This information is useful in estimating feeding capacity, adequacy of the sample, and when, related to time of collection, detection of periodicity in feeding.
3. Exact locality where collected and, on the next line, name of collector, collection number, date, and time of collection.
4. Refers to length of remains found in organ studied and this may be measured. When compared to sl. 5, below, it will give information on amount of digestion which has been progressed.
5. Refers to length when food organism was alive; for fish this is total length, but for many other animals one will record body length. Often the figure entered here must be estimated from fragmentary remains by comparing

the parts found with whole organisms or skeletons of specimens of known size.

6. The volume of food in ml may be given for each item as determined by water displacement employing a graduated cylinder with the smallest practicable diameter and the finest convenient graduations for greatest accuracy. Except in food studies of very small animals, record "Trace" for food items present in volumes less than 0.1 ml. in the next column to right give percentage of total volume of food made up by each item. Non-food items, such as, gravel may not be included in determining the total volume of food.

Food habits of _____

Based on the analysis of _____ ml. of food, the contents of _____ stomachs all of which contained 0.1 ml. or more of food (other stomachs are considered empty for practical reasons, at least in volumetric work).

Food Item	No. of Individuals of Each Food Item	Per cent Composition by Volume	Per cent Frequency of Occurrence

About 300 to 500 specimens, representing early post-larval stages to full grown adults are to be collected from different habitats during different months of the year to study their gut contents. As far as possible these specimens should be freshly caught, killed and immediately preserved on the spot so that the gut contents would not undergo any changes between the time of capture and preservation. Smaller specimens to be preserved as a whole, but the larger ones are examined first for relevant features, dissected and the gut taken out and preserved separately.

Broadly the gut contents are analyzed numerically, volumetrically and gravimetrically. The choice of method will depend upon the type of food the fish eat. Volumetric and occurrence methods should be adopted in all cases. Based on these methods, an index, known as "Index of preponderance" should be worked out.

Volumetric

1. Determine the volume of gut contents of the sample by water displacement method.

2. Sort the sample to kinds of items (species or larger groups).

3. Obtain volume of each kind of food item in the sample.
4. Compute percentage for each kind of food item that it forms in the total volume of food in the series.

Gravimetric

1. Obtain weight of gut contents in sample.
2. Sort the sample to kinds of items (species or larger groups).
3. Obtain weight of each kind of food item in the sample.
4. Compute percentage which each kind of food item forms in the total weight of gut contents in the series.

Numerical

1. Sort the gut contents to kinds of items (species or larger groups).
2. Count the individuals for each kind of food item in the sample.
3. Sum up each of them in the series.
4. Compute numerical percentage of each item in the gross total.

Occurrence Method

Method 1

1. Count the number of fish in which each food item occurs.
2. Express the above as percentage of the total number of fish examined.

Method 2

1. Sum up the occurrences of each food item.
2. Find the percentage of individual occurrences (kind of items).

In the case of volumetric and gravimetric analysis, a short cut method can be followed. Pool month wise the fishes, class-interval wise and indicate the analyzed data in terms of average size of the class interval. In the case of numerical analysis, take from the class-interval wise determined volume of gut contents, a few samples and determine the counts of each of the organisms. Compute the numbers per qualitative identity for the total volume.

Index of Preponderance

The volumetric, gravimetric and numerical methods of analysis emphasis only the quantitative aspects of gut contents, while "Occurrence" method indicate only the frequency of occurrence of food items. These methods individually are not suited for grading the food elements unless they are combined into an index. Index of preponderance is such a composite measure which takes both of quantity and occurrence into consideration simultaneously.

The state of digestion of gut contents indicate how far the food items in question are used in metabolism. For this purpose the rectal contents should also be studied and undigested food items recorded.

Chapter 11

Age and Growth Studies of Fish

Knowledge of age and of rate of growth in fish is extremely useful in management.

The following points reveal the practical application of age and growth work. A knowledge of age at attainment of sexual maturity is needed to properly regulate certain fisheries and to conduct certain fish cultural activities (how long will it be necessary to hold fish until they will of breeding age? If young are stocked, how long will it be before they will reproduce, *etc.*), age when catchable size is reached is important in regulation (how long will it take for a particular species to reach catchable size ?); determination of longevity may help recognition of environmental unsuitabilities; comparisons of rates of fish growth between bodies of water or under different environmental conditions may partly identify good and bad conditions and point the way for future actions; relation of age and growth in any one body of water to a regional average; helps measure environmental suitability for the species in question; a partial test of the success of attempts at environmental improvement is afforded by effect of the changes on growth rate; age and growth studies may also show suitability of stocking when used as a follow-up measure; continuing studies of age and growth in particular bodies of water will show the normal fluctuations from year to year and over period of years needed for the proper interpretation of deviations which single sample may show from a regional average (such long-range studies are currently in great need; characteristics of samples from one year have in general been too much relied upon as indicative of the growth characteristics of fishes in a given body of water).

Methods of Age and Growth Studies

There are several methods which have been used in age and growth studies in fishes; (a) length-frequency method; (b) known age method; (c) otolith and bone methods; (d) scale method.

Length-Frequency Method

This method is based on the expectancy that frequency analysis of the individuals of a species of any one age group collected on the same date will show variation around the mean length according to normal distribution (normal bell curve when graphed); it is further based on the theory that when data for a sample of the entire population is plotted, the normal curves for successive age-groups will be recognizable, thereby enabling their separation (Petersen, 1891).

There follow frequencies of various lengths in a series of pumpkin seed sunfish from a lake in August. For convenience these are arranged in 5 mm groups.

Length in 5 mm Groups	Number of Individuals	Length in 5 mm Groups	Number of Individuals
15	0	20	10
25	40	30	13
35	0	40	5
45	41	50	62
55	45	60	28
65	12	70	33
75	5	80	15
85	7	90	5
100	8	105	11
110	6	115	7
120	17	125	23
130	22	135	15
140	6	145	4
150	6	155	3
160	2	165	1
170	2	175	1
180	5	185	2
190	1	195	1

These data, when plotted as a graph using as length as the abscissa and frequency as the ordinate (frequency polygon), average lengths (modes) for each age-group may be obtained by interpolation of the tabular data from the graph. The age groups were determined by reading year marks on scales by a method proven valid for this species. Calculate the average lengths for each age-group for comparison of annuli on scales or other bony structures.

Size Frequency Analysis by Age Groups for Pumpkin Seeds

Age Group	Size Group (mm)	Frequency
0	15	1
	20	10
	25	38
	30	3
	35	1
I	40	5
	45	41
	50	73
	55	52
	60	30
	65	10
II	70	2
	75	5
	80	16
	85	8
	90	3
	95	2
	100	2
III	90	4
	95	3
	100	8
	105	12
	110	15
	115	7
	120	17
	125	22
	130	22
	135	15
	140	7
	145	4
	150	6
	155	2

Age Group	Size Group (mm)	Frequency
IV	155	2
	160	3
	165	1
V	180	5
VI	175	2
VII	185	2

Known Age Method

The known age method is applied when young-of-the-year fish is reared in the pond for one or more seasons and took periodic samples to determine the growth. This may also be done by releasing tagged, fin-clipped, or otherwise marked fish of known age in natural waters and recapturing them at intervals for study. Such procedures are of particular value in determining growth potential and in providing verification for other methods. They may serve, for example, as a means for identifying year marks on scales or other bony structures.

Otolith and Bone Methods

Otoliths (ear stones) and various bones such as ones in the opercular series, vertebral centra, ribs, scapulae *etc.*, show zones of differential deposition of bony material. When these marks can be correlated with annual growth, they are usable for aging fish. Otoliths often exhibit their growth zones best on one of their surfaces which has been polished on a fine abrasive wheel or stone. The otolith method has been used successfully (with limitations in older age-groups where the lines tend to merge) for sturgeon, for shad and plaice.

Age and Growth Studies by Scale Method

The principal types of scales of freshwater fishes are; ganoid, bony plates of sturgeon; rhombic, ganoid plates of gars; cycloid scales of trout, minnows, and most other soft-rayed fishes and ctenoid scales of perch, bass, sunfish and most other spiny-rayed fishes. With reference to cycloid scales, the focus, radii may be noted in addition the more or less concentric circuli (may be better called ridges on those scales where they are more or less longitudinal or transverse rather than concentric.

Scales may be partly or wholly imbedded (eel) or imbricated (overlapping like shingles). Embedded scales are variously shaped depending upon differences in direction of growth.

A part of the focus is the first part of the scale to be developed. It is often located at the center but in case of widely overlapping scales, its position is shifted due to unequal growth of the anterior and posterior areas of the scale. The direction of shifting may be ascertained in the scale of local minnow; the perch.

Ridges or circuli are produced on the first-formed stratum of the superior (outer) layer of the scale and are composed of a transparent homogeneous substance called hyalodentine. They are not the peripheral rims of superimposed laminae, plates or scalelets as has often been held and in fact are not deposited in any relation to them. According to Creaser, " a ridge is not built up simultaneously in all its parts. Various detached portions of its length may under construction at the same time". These parts may eventually unite to form a continuous circulus.

Seasonal cessation of growth or other factors may result in one or more discontinuous ridges on the scale becoming located between two continuous ones. This is one of the criteria which identifies an annulus or year mark. Scales having foci at their centers generally show ridges completely encircling the foci and annuli on such scales usually have the character just mentioned.

Scales in which the foci have shifted (sunfishes, minnows *etc.*) show some of the circuli ending at different places along the lateral margin at the time of annulus formation. On the resumption of growth, the new ridges parallel the entire scale margin and hence cut across the unfinished ends of the outcurved ridges. This feature of "cutting over" is a further aid in the identification of annuli of some fishes.

A third character very useful in the identification of the annuli of certain fishes is the relative approximation of the circuli. Circuli are often closer together just before the line which marks the annulus and farther apart just outside of it.

Sometimes the annulus is a clear, narrow streak encircling the focus (as the most prominent mark on the scale), but more often it is not so easily identified. The foregoing criteria aid in determining its approximate location.

Some scales will be found to have an abnormally large focus; these are replacement or regenerated scales. The enlarged focus represents the size of the scale lost by the fish. Replacement scales are hardly usable for age and growth work.

Validity of Annuli as Year Marks

In general, the use of scales, once proven applicable to a species, has shown itself to be the simplest and most accurate means of studying age and growth. Annuli on scales of many fishes are easily determined, after some experience and have been successfully used both in aging fish and in reconstructing their past growth history. The validity of annuli as year-marks has been proven for many species. The evidence used has been summarized by Hile (1941) on whose account the following comments have been based.

Annuli may be considered as year-marks provided;

1. There is correlation between the age of a fish and its size; provided (a) the regularity in increase in the number of annuli should be accompanied by increase in size of fish to prove that occurrence of annuli on scales is not haphazard but that annuli are added systematically as growth proceeds. (b) Modes in length frequency distribution of small fish should coincide with the modal lengths of corresponding age-groups based on scale readings.

2. There are agreements among calculated growth histories. (a) Lengths at the end of various years of life calculated from scale measurements

should agree well with the corresponding empirical lengths of younger age-groups whose ages were determined by the examination of scales. (b) There should be good agreement among data, (1) on calculated growth of fish of the same age-groups in different year's collections; (2) for different age-groups of the same or different year's collections; (3) on growth histories of the different age-groups (yearlings, two-year-olds, *etc.*) of the same year-class (a year-class is composed of the fish spawned in any one calendar year). (c) There should be good agreement among different year-classes as to the goodness or poorness of growth in certain calendar years.

3. There is persistent abundance or scarcity of certain year-classes in collections over several years.

The foregoing arguments are fundamental. They should be thoroughly understood. They may be applied to test whether growth marks are annuli and to determine whether or not evident annuli are valid. Validity of annulus-like markings should not be assumed for species where their worth has not been proven. The investigator should avoid being misled by "spawning checks", "false annuli" or other annulus-like markings which may be due to any one of the several causes, such as, resorption, interruption of growth by body injury, disease or spawning.

Collection of Scale Samples for Age and Growth Studies

A standard procedure is essential in the collection of scale samples for age and growth determinations. The utmost care must be exercised to gather complete data on all specimens and to be uniform methods of handling. An outline of methods in general use is given below. These are often modified to suit particular investigational needs.

Scale envelopes – Coin envelops are commonly used for holding scales and for recording field data. It is best to design headings with blank spaces for data on the envelop to cover all needs of an ordinary investigation, leaving space for additional notes.

To facilitate subsequent removal of scale samples, each envelop may be loaded with folded slip of paper into which the scales are placed. Before removing scales from the fish, it is better to complete the information called for on the envelop as follows:

Species – Common name and technical name after proper identification. Hybrids demand particular care for their identification.

Locality – Name of the lake, stream or river along with the locality where the fish was collected.

Method of capture – This information is useful in the interpretation of selectivity of gear and when the specimens are to be used for food study.

Time – Useful for determining periods of greatest activity and movements, and also useful in the study of feeding habits and food. The number assigned to the scale sample and written on the envelop may be repeated on a small piece of paper and slipped under the opercle or into the mouth of a specimen subsequently to be preserved as whole.

Name of collector and date on which collection was made.

Length – Ordinarily, lengths should be given to the nearest of one mm and should be taken when the fish is fresh and with its mouth closed and in a straight line, not over the curvature of the body.

Standard length – The most commonly used standard length measurement in fisheries operations does not always coincide with the standard length used by ichthyologists. The use of measuring board is responsible for this. The measurement will be different when the snout is terminal and when the lower jaw is terminal. The hidden bases of the caudal fin rays, where a groove forms naturally when the tail is bent from side to side is also to be measured.

Total length – This is the greatest possible length of the fish with the mouth of the fish closed and the caudal rays squeezed together to give the greatest over-all measurement.

Fork length – It is measured from anteriormost extremity to notch in tail fin of fork-tailed fishes (or to center of fin when tail is not forked).

Weight – The total weight to the nearest 0.1 gram as equipment permits or investigational needs demand.

Sex – The sex should be determined whenever possible. With some practice this can be done even with immature fish. It should be executed to the smallest possible sizes so that any shifts in sex ratio with age may be noted, and so that differences in rates of growth of the sexes may be determined. Sex should be determined by inspection of the gonads and not by observation of secondary sexual characters which often fail.

Maturity and development of sexual organs – It is necessary to determine that the fish is sexually immature, mature, ripe or spent. These factors bear on the coefficient of condition of the fish which may be calculated in the laboratory and on determinations of the length-weight relationship and should therefore be recorded at the time of capture.

To standardize the classification of stages of maturity of the sex organs, the following criteria may be adopted. During the spawning season of any particular species, classify the organs as immature, ripe or spent. By immature is meant that there are no grossly visible eggs or milt. Ripe means that the fish contains evident eggs or sperms and spent indicates that the fish has spawned.

Taking the Scale Sample

In order to be comparable, scale samples must be taken from the same region of the body. For some studies "key scales" are taken; these are identical scales or the same scale on each fish of the series as determined by count back along the lateral line and then so many above or below it.

In general investigations about 20 scales are taken from the fish. If it is a spiny-rayed form, they are taken from the side of the body just below the origin of the dorsal fin and just below the lateral line. In soft-rayed fishes, the sample is taken from the side of the body just above the lateral line and just below the insertion of the dorsal. In special investigations, it should be repeated, various other regions may be chosen.

If the fish is to be returned alive to the water, fewer scales should be taken, ten or even fewer.

The scales may be removed with a knife or with a forceps. Scraping the epidermis and mucous from the scales before removing them reduces subsequent labor required for cleaning at the time of mounting or making other preparation for study.

Preparation of Scales for Study

Several methods have been used for preparing the scales for study. Since magnification is necessary for reading scales, remains of soft integument, chromatophores *etc.* must be removed.

Cleaning scales – Scales may be cleaned in water by scrubbing them with a small hard-bristled brush or with a sharpened skewer. Soaking for an hour or more makes the process easier. Care must be taken not to break the margins of the scales and not to disrupt the softer, inner surface.

Temporary mounts – Wet or dry temporary mounts of cleaned scales may be made. Dry mounts are best when held between microscope slides; if the weight of slides does not keep the scales flat, the ends may be wrapped with rubber bands.

Permanent mounts – Such mounts are of two principal types : (1) celluloid or plastic impressions; (2) permanent mount of whole scales.

Impression method – Dry scales may be impressed on celluloid which has been softened by warming or by coating lightly with acetone. A convenient press may be made from a discarded notary's seal or jeweler's roller. The base may be electrically heated to warm the celluloid. Certain clear plastics are adaptable. Impression methods are particularly useful in preparing scales for study on which markings are obscured by opaqueness.

Mounting media methods – Easily obtained mounting media have been used for whole mount of fish scales are ; water glass (sodium silicate); glycerin-water glass mixture; karo syrup (clear, white), euparol; balsam; gum Arabic; polyvinyl alcohol (PVA); and glycerin-gelatin mixture. The latter mixture is at present most widely used because it is easily prepared, inexpensive and relatively permanent. It further has a reasonably suitable index of refraction and scales may be mounted directly from water. The slides, cover slips, and other equipment are easily cleaned with water for reuse.

The following formula (Van Oosten, 1929) gives good results – Dissolve 8 ounces of gelatin (sheet gelatin, white preferred, cut into bits) in 850 ml of water and add 250 ml of glycerin; warm the mixture to dissolve gelatin, taking care to avoid water loss. Place in small jars and add a few drops or crystals of phenol to each to keep molds from growing in them. Use hot (not boiling) for mounting.

Determination of Age and Rate of Growth in Fish

The assessment of age by reading annuli or year marks on scales has certain necessary preliminaries. The exact nature of the annuli and their identity as true year marks must first be established by tests described earlier. This may be done in part by applying the Petersen (length-frequency) method to a large series of specimens to determine size groups or annual increments of growth and comparing these with the marks on scales. It may also be done simply by checking annuli on scales from fish of known age.

Accurate assignment of age is often made difficult or impossible by lack of knowledge of time of annulus formation. Unless this time is known it may be difficult to tell whether the marginal area outside the last evident annulus represents the increment for the entire, previous season's growth or whether it constitutes growth during the new growing season. This is particularly true for fish collected in spring and early summer months. Study of scale samples taken systematically through the first six months of each year will answer the question for most species.

Photographic method – Scales may be used as photographic negatives and as such be projected onto enlarging paper (glossy, white, contrast) and studied in this way. Although providing a fine record of the scales, this method is costly and time consuming.

Microscopic method – An ordinary compound or binocular microscope may be used for assessing age of scales but measurements are rather precluded by this method. An ocular micrometer may, however, be used. For small scales a camera lucida may be employed to obtain the measurements needed for growth calculation or determination of body-scale relationship.

Scale projection method – Various types of micro-projection apparatus are in use for scale study. The basic type of machine is that of Van Oosten-Deason-Jobes (1935). In the construction of a scale machine any good micro-projector may be used provided that it will accommodate the required micro-tessar and other objectives.

Body-Scale Relationship and Growth History

The soundness of the scale method of determining the length of a fish at previous successive years of its life and its annual growth increments depends on the validity of the following assumptions (Van Oosten, 1929):

1. That the scales remain constant in number and retain their identity throughout the life of the fish.
2. That the annual increment in the length (or some other dimension which must then be used) of the scale maintains, throughout the life of the fish, a constant ratio with the annual increment in body length; and
3. That the annuli are formed yearly and at the same time each year (or that some other discoverable relation exists between their formation and increment of time).

As summarized by Van Oosten (1929), one may determine from scales; (1) the age of fish in years; (2) the approximate length attained by it at the end of each year of life; (3) its rate of growth for each year of life. It is now known how age is determined. The length at each year of life may be estimated by averaging the lengths of fish of the same age. The length of a fish at the end of each year of its life may be computed from a series of measurements of a scale when the length of the fish at time of capture is known. Given the total length of a scale, the length included in its annulus of year X, and the standard length of the fish from which the scale was taken, the standard length attained by the fish at the end of year X is determined simply by use of the following formula (Van Oosten, 1929), in which the third term is unknown (a straight line relationship of scale length to body length is assumed):

$$\frac{\text{Length of scale included in annulus of year}}{\text{Total length of scale}} \times \frac{\text{Length of fish at end of year}}{\text{Length of fish at time of capture}}.$$

Calculation of Fish Growth from Body Scale

Methods that have been used in the calculating of the growth of fishes from scale measurements given below as summarized by Ralph Hile. They are included here for the advanced scholars because of their importance.

The Dahl-Lea Method

This method assumes that the mathematical relationship between the body length and scale length is expressed by the equation;

L = cS

where;

L = Body length (standard length);

S = Scale length (refer to the dimension of the scale actually measured in the study of the growth of each particular species. The anterior radius is most frequently measured, but some investigators have employed the posterior radius and some have measured the diameters of the different growth areas)

c = A constant

The Dahl-Lea method of growth calculation therefore holds that the ratio of body length to scale length is constant for all lengths of the fish beyond that at which the first annulus is laid down.

The Lee Method

This method assumes that the mathematical relationship between body length and scale length is expressed by the equation,

L = a + cS

L, S, and c have been defined earlier; a is also a constant. Here the body – scale ratio (L/S) varies between the limits of infinity as L approaches and c as L increases without limit, but the ratio of the corresponding increments of body length and scale length is constant. Frequently attempts have been made to interpret the length of the intercept, a, as the body length at which scales first appear on the fish. In some species this interpretation may be approximately correct, but it cannot be accepted as a generalization, since the intercept may be negative in some species (Monastyrsky).

The Sherriff Method

This method assumes that the mathematical relationship between body length and scale length is expressed by the equation,

L = a + bS + cS square,

Where, a, b, and c are empirically determined constants. On purely mathematical grounds it is apparent that Sherriff's equation must fit a mass of empirical data on body length and scale length better than Lee's linear equation. This method has been little used. Monastyrsky applied the Sherriff equation to several species of fish and held that the method is open to the same general criticisms as the Lee method, particularly in the interpretation of the constant a, as the length of fish at which the scale first appears.

The Jarvi Method of Arbitrary Correction

An arbitrary correction is made by adding a given quantity (determined empirically) to the calculated length for each year that intervened between the time of capture and the year of life for which the length calculation was made. For example, 5, 10, 15, or 20 mm would be added if 1, 2, 3, or 4 years, respectively, intervened.

The Monastyrsky Logarithmic Method

This method holds that the logarithms of fish length and scale length exhibit a straight-line relationship, or that,

Log L = log c + n log S

Monastyrsky applied the method to several species of fish and found the calculated growths to agree well with the observed growths. He held the results obtained from his logarithmic method were far superior to those obtained on the same species by the Lee or Sherriff methods. Monastyrsky made his calculations by means of a logarithmic nomograph.

The Segerstrale Empirical Body-Scale Relationship Method

In this method the average scale lengths corresponding to different body lengths are determine through an extensive series of measurements of key scales taken from selected area of the body. The resultant body-scale relationship expressed in tabular form, or as a curve, then serves as the basis for the calculation of the growth histories of individual fish. In his publication he described a mechanical device for the calculation of growth from the curve of the empirical body-scale relationship.

Fry's Modification of the Monastyrsky Method

Fry added a constant to the Monastyrsky equation to give it in the following form;

Log (L-a) = log c + n log S

Fry's belief that the constant a corresponds to the length at scale formation is of course subject to the same criticism outlined previously. The introduction of the additional constant created the difficulty that a mathematical fitting of the equation is impractical. In practice different values of a must be tried, the one giving the closest fit determined by inspection, and the values of n and c estimated from the graph. Fry described a nomograph for the solution of his question. Although the nomograph is satisfactory, it should not be used as directed.

Carlander's Third Degree Polynomial Method

Because of the sigmoid character of his graphical representation of his empirical data, Carlander employed a cubic of the following type;

L = a + bS + cS square + dS cube

He calculated growth histories of individual fish from a table following the same general procedure used by Segerstrale.

On purely theoretical grounds the Segerstrale method appears to be the most satisfactory of those listed above, first because it is based on a detailed examination of the actual size of the scale at different body lengths and second because it involves no assumptions of a fixed mathematical relationship between the body length and scale length. The method is faced, however, with the practical difficulty of determining a completely reliable empirical body-scale relationship for the complete range of body lengths over which growth calculations are to be made. The length distribution of a fish population is ordinarily of such a nature that it is difficult or even impossible to obtain adequate representation of certain lengths. This difficulty is particularly great for those species in which the young fish cannot readily be located and captured. Weaknesses in the empirical body-scale curve resulting from inadequate sampling at certain lengths must inevitably be reflected in accuracies in growth calculation.

The later methods of growth calculation may all be considered as attempts to improve on the first (Dahl-Lea method). Each has led to more accurate growth calculations for the species to which they are applied. However, the fact that different authors, working with apparently reliable material, have arrived at such a diversity of conclusions concerning the nature of methods of calculating growth suggests that the question of the body-scale relationship may not be subject to generalization. The body-scale relationship and growth calculation problems should be considered specific rather than general. Almost certainly the nature of the body-scale relationships varies from one species to another, and it is not improbable that the retationship may vary between races and populations of the same species.

Length-Weight Relationship

Weight in fishes may be considered a function of the length. If form and specific gravity were constant throughout life the relationship could be expressed by the well known cube law in the equation; where,

W = KL cube, W = weight, L = length and K = a constant

Actually, in nature, it has been found that the value of K is not constant for an individual, a species, or a population but that it is subject to a wide range of variation.

The values of K under various definitions (coefficient of condition, condition factor, length-weight factor) have been used widely by fishery investigators to express the condition of relative robustness, "degree of well-being" of fishes. On this basis K has also been used as an adjunct to age and growth studies for indicating the suitability or lack of suitability of an environment for a species. By comparison of the value for a specific locality with that (an average) for a region. It has been employed to measure the effects of environmental improvement.

The investigator may expect to find the following characteristics of K values;

1. Values increase with age. In other words, in later life fish gain proportionately more in weight per unit of length increase than they do in earlier life.

2. Sexual dimorphism is often exhibited in K values. One must not assume absence of such a difference but must test to see if it is present. If present and sufficiently great, the sexes should be kept separate when computing mean values.

3. Sharp changes in K values may be expected at spawning. Consider the value of K just prior to spawning with that for the same fish right after spawning.

For best comparisons, values for K which are to be compared should be for fish of the same length, age, and sex and the fish should have been collected from the different waters on the same date. Obviously, this ideal cannot be realized in practice. But most investigators strive to base comparisons at least on same sex, length and age for fish collected in the same season of the year. If comparison of values for K is to be made between individuals from a single body of water, the collection should be made as nearly as one date if possible and it should be carried out with precautions against the effect of selectivity of the collection method (gear plus mode of operating gear).

But the foregoing equation has also been employed to describe the general length-weight relationship in populations of fishes, and thus to serve as the basis for the calculation of unknown weights of fish of known length or to determine the lengths (unknown) of fish of known weight. The use of the equation in this latter capacity has met with indifferent success, due to the failure of the cube law on which it is based to describe accurately the relationship of length to weight in many forms of fishes (for example, some fishes increase somewhat more in other dimensions than in length since their weight at a given length is greater than the weight calculated from the law).

The instances where the cube law does appear to apply to the length-weight relationship apparently are the exceptions and are coincidences. The following more general equation is a much more satisfactory method of describing this relationship in fish :

$$W = CL \, n,$$

Where W = weight; L = length and C = a constant

In this equation the values of both C and n are determined empirically, that is they are derived from data taken directly from specimens in large series.

It may be concluded, therefore, that the description of condition and the expression of the length-weight relationship are two entirely different things. Coefficients calculated from the cube relationship and those from empirically determined exponents are in no sense of parallel significance as measures of condition.

In fishery management it is often very useful to be able to determine the weight of a fish when length alone is known (or vice versa). To provide a basis for such estimates, an empirical curve may be fitted by inspection to points plotted from lengths and corresponding weights.

The nature of such a curve may also be computed mathematically from length-weight data. From graphs of such curves length may be estimated when the weight alone is known or vice versa. If the mathematical relationship between length and weight has been formulated, the formula giving the best fit may be used to calculate a corresponding length or weight where one or the other is known.

Length-weight knowledge may also be useful in regulating fisheries. If for example, there is no market value for a fish unless it weighs one-half kg or more, and if length for this weight could be estimated from an existing curve, one may set the mesh size of the gear to provide escapement of smaller individuals without additional, time-consuming field work. They may then grow to the desired weight before being taken.

Chapter 12

Fish Population Census

A knowledge of the magnitude and composition of fish populations is very useful in management. Fish population studies have a bearing on (1) production per unit area- by numbers, by kinds and by weight of fish; (2) appraisal of good and poor species combinations; (3) along with age and growth studies and catch statistics, effects of fishing and fishery regulations. No simple method of estimating fish population existed, even the tools which have been developed are not being widely used.

The methods of fish population study in streams have been mostly to block off sample sections and to remove fish by poisoning, by seining or by electrical fishing. The mark and recapture method has also been used with varying degree of success in both flowing and standing waters. It is apparent that if certain requirements are met, this method can give reasonably reliable estimates for certain water areas. Owing to differences in response to netting among species, any method which involves the use of such gear will give better results for some species than for others. Some attempts have been made to base estimates of fish populations, or at least, estimates of population density on catch per unit of gear but individuality of species response and seasonal variations seem to make the general results indifferent. In the commercial fisheries, however, catch per unit of effort, properly adjusted, is an important means for detecting fluctuations in abundance.

Several methods have been described for handling the mark-and-recapture method of estimating fish populations. In the simplest one, a known number of marked fish of the species in question is placed in the body of water. Subsequent collections yield some marked and some unmarked fish. The simple proportions applied to the catch for a given unit of time or effort gives population estimates which may be averaged.

$$\frac{\text{No. of marked fish recaptured}}{\text{Total number of fish captured}} = \frac{\text{Total no. of marked fish in lake}}{\text{Total population in lake}}$$

Such estimates obviously include only those sizes selected by the gear.

A modification of Schnabel's (1938) method has been developed by Thompson in Illinois and has been widely used. This method weights daily averages as it proceeds. On each of several successive days of capturing, marking, and releasing fishes, the population is estimated by the following formula:

P = Summation AB/Summation C where,

P is the estimated population on any date,

A is the number of fish caught on any day.

B is the number of marked fish present in the lake on any day,

C is the number of returns on any day,

AB is the product of A and B,

Summation AB is the sum of all products (AB) calculated to date,

Summation C is the sum of all returns to date.

Another method possibly applicable to the estimation of fish populations has been given by Du Lury (1947). This, the graphical intercept method, may be used in situations where fish can be removed at such a rate that each removal reduces the number that will be taken in succeeding trials. To estimate a population by this method, one makes a graph of the number of fish taken per trial (ordinate) against total catch to date (abscissa). In other words, catch per unit of effort (by individual trial, by day, or by week *etc.*) is plotted as the ordinate against the cumulative total catch, the abscissa. After several points have been found, they are fitted (usually by a straight line). Extrapolation of the fitted line to an intersection with the abscissa will give an estimate of the total population at the intercept.

These population calculations depend on certain assumptions. It is assumed, that marked fish are normal in behavior and just as viable as are wild, untrapped stock. There is some evidence that this condition is not strictly met since dead marked fish are sometimes found floating on the lake. This will tend to make estimates too great. Releasing fish near the traps in which they were captured might cause them to reappear more frequently than normal and this would counteract effects of mortality. Another source of error is natural mortality combined with the growing of young fish into the legal-size group (recruitment from the smaller size groups).

Field Study

If suitable situation is available locally, a small population study in the field can be conducted. A small easily seinable pool or rearing pond with a few hundred of fish in it will be sufficient. Instead of conducting this work in the field, it may be done experimentally in the laboratory in the following manner.

1. Set up the data sheet showing columns (I) no. of beans taken each time, (ii) no. of beans marked and returned, (iii) no. of marked bean present in the pan on a time (iv) the product of (i) and (iii), (v) sum of all data of (iv), (vi) no. of bean returned each time, (vii) sum of all returns and (viii) estimated no of bean in the pan.

2. Place a few hands-full of dried beans (the "fish") in a pan (the "lake").

3. With eyes closed to ensure a random sample remove a few fingers full.

4. Count the seeds, mark each with a pencil or ink, record numbers in proper place on the data sheet, return to pan, and mix by stirring thoroughly.

5. With eyes closed again, take another like sample, count, record marked and un-marked ones, and return all to pan, mix calculate total number in pan.

6. Repeat (5) until begin to approximate one another.

7. When estimates become more or less constant, verify estimate last obtained (presumably the best of the series) by actually counting all seeds in the pan. In counting, separate the marked and unmarked seeds in order to verify previous counts of marked ones. State error of estimate in per cent and consider the following requirements of fish population studies by the mark and recapture method in the light of the deviation noted.

Some requirements which should be considered in fish population studies by the mark and recapture method are;

(a) The work should be conducted intensively over a short period of time in order to reduce errors from mortality and from recruitment by growth of smaller fishes into the size ranges being handled.

(b) Fishing effort is best when more or less regularly distributed over the whole body of water to give more representative sampling.

(c) A basic assumption is that marked fishes mix with unmarked ones freely and do not shy away from nets as a result of having been handled once or are not attracted to recapture.

(d) A large number of the fishes must be marked to give reliable results; this may be as great as 60 per cent of the total population but in most studies such a high percentage would be impractical.

(e) Handling and marking of the fishes must not make their chances of survival less than for those not handled during the course of the study.

Chapter 13

Fish Diseases

Like all animals, fish have a full complement of diseases, parasites, and abnormalities, both malignant and benign. There is no question that most fish die from such disorders, natural enemies other than man, or old age- certainly not from being caught by fishermen. Usually the ailments of fish may be placed in one of the following four categories; (1) disorders resulting from external physical and chemical agencies, such as, mechanical injuries or pollution; (2) dietary or developmental deficiencies, such as, stunting and certain types of malformation, (3) tumors and carcinomas; (4) diseases including parasitization. Studies of these disorders constitute the field of fish pathology, with etiological, morphological, physiological and therapeutic aspects.

Diseases are one of the most important problems in aquaculture. Disease is any alteration of the fish body or one of its organs so as to disturb normal physiological function. All species of fish have characteristic protection, food-gathering and spawning behavior. In general, gathering at the surface or pond edges can often be a sign of disease.

All fish do not get sick and die each time when a disease outbreak occurs. There are many factors, which affect how an individual fish responds to a potential pathogen. The pathogen must be capable of causing disease. The fish must be in a susceptible state, and certain environmental conditions must be present for a disease outbreak to occur. The following behavioral, changes in appearance and clinical signs would help to determine whether or not the fish is diseased.

Behavioral Changes

☆ Off feed

☆ Lethargy and isolation

☆ Opercular flaring

☆ Changes in respiratory rate

☆ Hanging listlessly in water

☆ Rubbing against objects

☆ Gasping at surface Clamped fins

☆ Flashing

☆ Equilibrium disturbances

Changes in Appearance

☆ Skin lesions-nodules, redness, ulcers

☆ Scale loss

☆ Changes in color

☆ Bulged eye- exophthalmia or pop-eye

☆ Sunken eye- enophthalmia

☆ Ascites and distended abdomen

☆ Emaciation

☆ Abnormal body confirmation

☆ Improper buoyancy

☆ Death

First-Aid for Diseased Fish

☆ Identify original pond with diseased fish

☆ Isolate individuals into smaller tanks

☆ Evaluate water quality records for acute/chronic aberrations or trends

☆ Check the water quality for dissolved oxygen, ammonia, nitrite, and pH

☆ Correct the basic problem, usually a management or an environmental problem

☆ Work up fish with general diagnostics

☆ Do not blame disease for mass mortality

☆ Do not turn to chemicals to solve problems.

☆ Chemicals degrade the pond bottom, water quality, affect fish health and plankton bloom.

☆ Contact fish health professionals for diagnosis and remedial measures.

Common Diseases

Fish may contact diseases from which they are required to be protected. Injuries can be caused during handling, which may break down resistance to numerous parasitic diseases and physiological disorders. Intensity of population, malnutrition and similar factors may also prove harmful to fish stocks. Well kept and properly managed ponds usually contain healthy and thriving fish populations, provided the fry and fingerlings stocked were originally in good condition. Carelessness in stocking and feeding may result in serious parasitism, debility and subsequent mortality. Even though several curative methods are available, treatment is difficult and often impracticable in ponds containing large numbers of fish. Prevention is the most effective method of control. Some of the common diseases of fish are given below.

Fungal Infection

Saprolegnia parasitica, the most common fungus attack eggs, fry, fingerlings, and adults, when they get injured either mechanically or as a result of parasitic infections. Red patches on the body are observed in infected fishes.

The popular treatment are baths in (a) 3 per cent common salt solution, (b) 1 : 2000 or 1 : 3000 copper sulphate solution, (c) 1 : 1000 potassium permanganate solution, till the fish shows signs of distress.

Gill rot is caused by the fungus *Branchiomyces* spp. Among cultivated fishes in ponds having abundant decaying organic matter, the gill filaments, particularly the upper ones become blackish red in color and the rest whitish, as a result of penetration of fungus which obstruct the blood vessels.

Bacterial Infection

Fin rot or tail rot is caused by a rod shaped bacterium, which grows on fins. The fin loses its rays and reduced to stump. The disease is fatal to younger fish. The best treatment of fin rot is a bath for 1-2 minutes in a 1 : 2000 solution of copper sulphate by a skilled person, who can adjust the period of treatment according to the behavior of the fish. The treatment should be stopped as soon as fish shows signs of distress.

A bacterial ulcer (ulcer disease) affecting the major carps, where open sores or ulcers on the body of the fish increase in size gradually exposing muscles, can be treated by one minute bath in 1 :2000 copper sulphate solution for 3-4 days.

Infectious dropsy among cultured major carps in India has been reported in late winter. Thorough disinfection with 1 ppm potassium permanganate solution or dip treatment in 5 ppm of the same chemical have been found effective in checking spread of disease.

An epidemic eye disease, caused by the bacterium, *Aeromonas liquefaciens* affecting medium and large sized catla can be cured in early stages of infection with hourly baths in 8-10 mg/liter chloromycetin solution for three consecutive days. The infection affects the eyes, optic nerves and brain of the fish and cause large scale mortality.

White spot disease, the commonest protozoan disease is observed in freshwater fishes. Fish mortality due to this infection have been observed in nursery and rearing ponds. *Trichodina indica* and *Scyphidia pyriformis* are the parasites causing this disease in carps in India. A bath in 2-3 per cent common salt solution for 5-10 minutes will be useful.

Common mastigophoran infection (Costiasis) caused by *Costia necatrix* and *Bodomonas rebae* can be controlled by a bath for 5-10 minutes in 3 per cent common salt solution.

Parasitic Infection

Several species of *Myosporidia* produce cysts on different regions of the body and in internal tissues and organs and sometimes causes very difficult situations for fish culturists. Nothing much is known for effective cure. Among monogenetic trematodes, dactylogyrids and gyrodactylids cause serious infections by fading of colors, dropping of scales, excessive mucus on caudal peduncle and fins and peeling of skin. During epidemics, badly affected fishes should be destroyed and mild infections be treated for half an hour in 1:100000 potassium permanganate solution or 1:2500 formalin.

Tape worms infected fishes caused by cestodes appear dull, sickly and with parts of alimentary canal swollen or completely chocked by hundreds of cestode cysts. No effective control method is known so far.

The infected fish by argulus (argulosis) become very weak and emaciated with stunted growth, loss of scales and red spots in sites of puncture. For control, draining the water and exposure of bottom soil to sun at least for 24 hours is desirable.

Viral Diseases of Fish

Among diseases, viral infections in culture environment have resulted a decreased fish production and income. At present, viral diseases are less known in warm water fishes due to lack of diagnostic techniques. To some extent the problem has been overcome by the application of fish cell lines in tropical countries.

Viruses are submicroscopic particles that multiply only within the living cells of an animal or plant host. Other micro-organisms, such as, bacteria, fungi have organelles for their own metabolism, but viruses do not. They must utilize the machinery of the infected host cell for growth and reproduction. A virus has two parts, the internal part, the virion or virus particle, which is composed of nucleic acid, the same material that makes up genes. The virion is enclosed in an external protein coat, called a capsid. Viruses are broadly categorized by the type of nucleic acid they contain. The two basic types of nucleic acid are RNA (ribonucleic acid) and DNA (deoxyribo-nucleic acid). Virologists also classify viruses by their shape, for example, "icosahedral" viruses have 20 sides and "helical" particles are rod shaped.

Being very small, viruses are often difficult to detect. Three techniques are used for initial identification of a virus. First electron microscopy (EM) is used to visualize virus particles within tissue cells. Second, an effort is made to grow the virus in the laboratory using established cell-lines, which are living cells grown *in vitro*, literally "in a glass" outside of a living organism by feeding them special nutrients. This technique is referred to as cell culture, and cells from specific fish are used for growth of specific viral agents. Finally identification of virus is confirmed using serology, in which serum (part of the blood) from animals known to be infected with the virus is tested for its ability to "recognize" the suspected virus.

This confirms that the virus in the animal's body is the same as the virus that has been isolated in the laboratory.

Viruses are often both species-specific and tissue-specific. This means that they may only grow in certain type of cells from certain animals. This can make it difficult to isolate viral agents from many fish because there may not be a commercially available cell line for an individual fish species. Many cell lines which are commercially available originate from cold water fish, such as, salmonids and may be less suitable for warm water species. It is impossible to develop serology as a tool until after the virus has been isolated in the laboratory using a suitable cell line. For these reasons, viral agents of many fish are often suspected based on visualization of viral particles in the tissues taken from sick fish using electron microscopy. The problem with this tool, when used alone, is that it is possible for viral particles to be present in tissue without causing harm or disease. Therefore, identification of viral particles in tissues of sick fish does not prove that the observed virus is the cause of the disease in progress.

Other advanced antibody based technique like ELISA, immunodot and western blot are also used in viral disease diagnosis. Besides, immuno-histo-chemistry (ICH), immunoperoxidase (IPO) and immunoflourescence (IFT) give detailed information about the target organs and cells involved. At present nucleic acid based techniques like PCR&DNA hybridization are becoming popular in diagnosis tools, owing to the simplicity of operation, less time and more accuracy.

Koi Mass Mortality

Causitive agent – Koi Herpes virus (KHV)

Host – Common carp (*Cyprinus carpio*) and koi (*Cyprinus carpio koi*)

Clinical Symptoms

1. High mortalities ranging from 80 to 100 per cent,
2. Lethargy with sluggish and erratic swimming,
3. White patches on the gills caused by gill necrosis and excessive mucus production
4. White or pale patches on the body with ulceration.
5. Fin rot, and
6. Enlargement of the kidney and liver with haemorrhages and discoloration.

Transmission - Horizontal

Treatment

1. Water temperature control.
2. Vaccination with attenuated non-pathogenic viruses, and
3. Transport restriction from known infected areas.

Lymphocystis Disease

Iridovirus is the causative agent. Most fresh and salt water species are observed to be the host. Clinically, fish are seen with variable sized white to yellow cauliflower-like growths on the skins, fins and occasionally on gills. Occasionally this virus may go systemic with nodules on the mesentery and peritoneum. Nodules may last for several months and cause infected fish to be susceptible to secondary bacterial infections. Re-infection can occur.

Histopathology – Fibroblast undergoes cytomegaly with many basophilic cytoplasmic inclusion bodies and a thick outer hyaline capsule. The inflammatory response is variable, but is usually a chronic lymphocytic inflammatory infiltrate. The disease gains entry through epidermal abrasions. The virus infects dermal fibroblasts. The disease is self limiting and refractory to treatment (screen out the infected fishes as soon as possible to prevent cross contamination).

Herpesvirus Salmonids

The causative agent being Herpesvirus, the disease is primarily observed in the fry of rainbow trout. Clinically the fish are lethargic with prominent gill pallor. Mucoid faecal casts are commonly observed trailing from the vent.

Lesions – Exophthalmus and ascites. Low hematocrit and numerous immature erythrocytes and haemorrhage in eyes and base of fins.

Histopathology - Histopathologically, multifocal areas of necrosis of the myocardium, liver, kidney and posterior gut. Syncytial cells involving the acinar cells of the pancreas are considered to be a pathognomonic sign. Transmission of the virus is believed to be direct. The disease can be controlled by avoiding exposing susceptible trout to the virus. If the disease occurs, raising the water temperature to 15 degree Celsius or more will minimize losses.

Channel Catfish Virus

The disease is observed in fry or fingerlings of channel catfish (less than 10 g weight) during the summer, when water temperatures are above 22 degree Celsius. The causative agent of the disease is Herpesvirus. Clinically infected fish usually show erratic swimming or spiraling followed by terminal lethargy. Mortality is very high. Lesions and haemorrhages at the base of fins and skins. Ascites, exophthalmus and pale gills. Kidneys are swollen and pale with haemorrhage. Spleen is enlarged and dark red and gill are usually pale.

Histopathology – Histopathologically, multifocal areas of necrosis and haemorrhage are observed in the posterior kidney, liver, intestine and spleen.

Infection is direct with transmission of the virus in the water or feed. Piscivorous birds, snakes or turtles may mechanically carry the virus from pond to pond. Survivors are persistently infected and become carriers for life. Control of disease is by sanitation, purchasing of virus free brood stock and lowering the water temperature to less than 19 degree Celsius during an outbreak to lessen the mortality.

Fish Pox (*Epithelioma papillosum*)

Non-fatal disease, observed in carp and other cyprinids, causative agent being *Herpesvirus cyprinid*. Elevation of epidermis with the formation of white to yellow plaques over the body of the fish. Healed lesions usually turn black. There is epidermal hyperplasia with the epithelial cells occasionally demonstrating intranuclear inclusion bodies. Transmission is probably direct. Treatment consists of rearing fry and fingerlings in virus free water, certification, notification of infections and zoning of disease-free and infected areas and hygiene controls.

Infectious Haematopoietic Necrosis (INH)

The disease caused by Rhabdovirus, is observed in the fry of rainbow trout and salmon (Chinook and Sockeye) with 100 per cent mortality. Fish become lethargic or hyperactive and dark due to increase in pigmentation. Exophthalmus, abdominal distension and faecal cast. Haemorrhage on skin and viscera, primarily at the base of fins, behind skulls and above the lateral line. Anaemia with pale gills and surviving fish may develop scoliosis. There is prominent necrosis of haematopoietic tissue including melanomacrophages of the kidney, red pulp of the spleen and hepatic parenchyma. Necrosis of the submucosal eosinophilic granular cells is considered pathognomonic for IHN. Intranuclear and intra-cytoplasmic inclusions are occasionally observed in acinar and islet cells of pancreas. The virus is transmitted by direct contact with infected survivors or by feeding contaminated feed. The virus is probably shed in contaminated semen and eggs. The disease is most severe at 10 degree Celsius and rare in temperature above 15 degree Celsius. Treatment consists of disinfection of eggs, rearing fry and fingerlings in virus free water and certification and notifications of infections and zoning of disease free and infected areas.

Viral Haemorrhagic Septicaemia

Being caused by Rhabdovirus, the disease is wide spread and very contagious in rainbow trout. The disease occurs in temperatures below 14 degree Celsius. Turbot, sea bass and Atlantic salmon are commonly affected by similar viruses. In acute disease high mortality occur in affected fish. Fish have pale gills, dark body color, ascites, exophthalmus and erratic swimming behavior (spiraling). Haemorrhage is common findings in the eyes, skin, serosal surfaces of intestines and muscles. Necrosis of the haematopoietic and lymphoid elements of the anterior kidney and congestion and necrosis of the hepatic parenchyma are histopathologic findings. A slower prolonged mortality occurs in chronic disease. Fishes become lethargic, have pale anaemic gills, drak skin, exophthalmus and distension of abdominal cavity. Internal organs are commonly involved with splenomegaly, hepatomegaly and swollen kidneys. Transmission is believed to be direct with contact of carriers, contaminated water and feed. Stamping out (infected farms drained, disinfected and sludge is removed prior to restocking with disease free stocks) program to be adopted alongwith preventing movements of birds, animals and infected water or fish into the farm.

Infectious Pancreatic Necrosis (IPN)

Most salmonids, primarily rainbow trout and brook trout are affected by Birnafvirus. The disease is characterized by sudden explosive outbreak with high mortality. Affected fish become dark and rotate their bodies while swimming. Diseased fish have distended abdomens and exophthalmus. The presence of a gelatinous material in the stomach and anterior intestine is highly suggestive of IPN. Mucoid faecal cast are common. Infected fish commonly have a low hematocrit and haemorrhage in gut, primarily in the area of pyloric caeca. Necrosis of the pancreatic acini gut mucosa and renal haematopoietic elements. A moderate inflammatory infiltrate is usually observed around pancreatic acini. Hyalin degeneration of skeletal muscle is also observed. Virus can be transmitted vertically in the eggs. Control methods include the screening of fertilized eggs and administration of a recombinant vaccine together with bactrins against vibriosis and furunculosis as polyvalent vaccine.

Spring Viremia of Carp and Swim Bladder Infection Virus

The disease occurs in carps and other cyprinids by several sub-types of *Rhabdovirus carpio*. Clinical symptoms consists of loss of coordination and equilibrium. Exphthalmus and abdominal distension (ascites). Inflammed and swollen vent. Oedema and haemorrhage in many organs. Pronounced inflammation and haemorrhage of swim bladder. Virus is shed in faeces and found in contaminated eggs. Treatments include hygiene controls, by heating to 45 degree Celsius, formalin, sodium hydroxide and chlorine. Certification and declaration of disease free zone is necessary.

Viral diseases cannot be controlled with medication, because they use the host's own cells for reproduction and survival. A good nursing care to be provided for fish suspected of having a viral infection, so that their own natural defence mechanisms can work to eliminate infected cells. This involves maintaining of excellent water quality, feeding fish with high quality diet, maintaining clean facilities and keeping sick or potentially infected stock separate from all other fishes.

Chapter 14

Pollution of Water Bodies

The problems of river pollution by industrial wastes and domestic sewage are challenging the best efforts of administrative, legislative and research bodies to effect an economical and rational balance between sensible regulation and industrial and urban expansion. It thus becomes necessary for any fish manager to recognize pollution and be able to recommend practicable means for its avoidance.

As regards the water purity standards for fish life, " general water conditions favorable to, not merely sublethal for, mixed faunae of fishes of the " warm-water" types and supporting organisms, present a complex defined by:

1. Dissolved oxygen not less than 5 ppm.
2. pH range between 7.0 and 8.5
3. Ionizable salts as indicated by a conductivity between 150 and 500 mho x 10 to the power minus six at 25 degree Celsius and in general not exceeding 1000 mho x 10 to the power minus six at 25 degree Celsius.
4. Ammonia not exceeding 1.5 ppm.
5. Suspend solids of a hardness of one or greater, so finely divided that they pass through a 1000-mesh (to the inch) screen; and so diluted that the resultant turbidity would not reduce the millionth intensity depth for light penetration to less than 5 meters.

" If such favorable conditions for fishes are to be maintained and fishes and other aquatic organisms are to be protected against the toxic actions of many river pollutants, all pollutants not readily oxidizable or removable by the river should be excluded including particularly all cellulose pulps, wastes carrying heavy metallic ions and gas factory effluents. Other types of wastes should be diluted to concentrations nontoxic to the aquatic life of the particular water body. No substance should be added to stream waters which could cause a deviation in general conditions beyond the limits outlined above."

Local sites of industrial pollution may be visited and the nature and extent of pollutants may be ascertained using the following blank form :

The means of detecting pollution are physical (odor, color, turbidity, *etc.*), chemical (analysis of dissolved substances, oxygen deficiencies, *etc.*) and biological by(a) index organisms (*B. coli, Tubifex* worms, pollution fungus, *etc.*) (b) test organisms, or by (c) determination of biological oxygen demand (for certain types of organic effluents).

For the situation investigated, experiments to determine the effect of polluting substances on fish and tolerance of fish to various concentrations may be determined. For treatment of domestic sewage, notes on the process for the treatment of domestic sewage adopted may be taken by using the following blank form.

Industrial Pollution Analysis Form

Location _____

Polluting agency (and responsible person) _____

Nature of pollution and how detected _____

Continual _____ Intermittent _____

Physical evidence _____

Chemical evidence _____

Biological evidence _____

Specific pollutants _____

Reliable limits of toleration for fish _____

Toxicity to aquatic life;

At point of effluence (include measurement of amount and concentration of effluent and volume of water flow) _____

Extent of toxic zone (compute from dilution) _____

Zone of recovery (compute from dilution) _____

Damage of pollutant (monetary value)

Esthetic _____

Recreational _____

On fish *etc.* _____

Abatement of pollution

What is being done, if anything ? _____

Recommendation for and description of control _____

Date _____ Analyst _____

Domestic Sewage Treatment Analysis Form

Location of plant _____

Community served _____ Population _____

Plant superintendent _____

Date of construction _____ Rated capacity _____

Volume _____ Expected life _____ Years _____

Flow in liters;

Peak _____ and time of day _____

Minimum _____ and time of day _____

Average _____ and time of day _____

Excess of peak over rated capacity _____

Location and number of lift or pumping stations _____

Primary treatment (clarification of sewage, describe)

Screening and fate of screened material _____

Grinding and fate of ground material _____

Primary settling and fate of settled materials _____

Secondary treatment (processing the liquid from primary, description)

Filter process (trickling, high rate, *etc.*) _____

Activated sludge process _____

Final disposal of liquid (where does it go? Is it chlorinated ? what is its B.O.D. ?)
_____

Treatment of residual sludge from primary and secondary treatments

Digestion _____

Gas from digestion (use?) _____

Drying final residue of sludge (air, vacuum filtering, centrifugation,
heat ?)_____

Final disposal of sludge (how treated for fill, fertilizer or incineration) _____

Number of employees and classification _____

Annual budget and breakdown of operating costs _____

Plant deficiencies _____

Adequacy of treatment in regard to use of water into which plant discharges _____

Recommendations for plant improvement _____

Date _____ Analyst _____

Chapter 15

Biological Fishery Surveys

Biological fishery surveys are investigations of fish populations and the chemical, physical and biological factors affecting them. The units of study are individual streams, lakes, ponds or impoundments. Often all of the waters in a geographical regions are explored, sometimes an entire, extensive watershed over a season.

The purposes of fishery surveys are; (a) to obtain fundamental information on fishery resources and conditions of the environment needed to develop befitting conservation policies and management practices; (b) to lead to closer integration of research and management programs of the various agencies concerned in fish and wildlife preservation and utilization in the region. Such surveys and related studies contribute information on the kinds and distribution of fishes in an area and on the factors leading to depletion of the fisheries. Properly applied, this information may bring about development on a higher level of water use for fishing and other forms of recreation, and for power, industry, transportation, and domestic consumption.

Data from biological fishery surveys form the basis for fishery management plans. Wasteful, or even harmful stocking is avoided by determining kinds and numbers of fish to stock. Need for aquatic plant introduction or control is identified. Overpopulations are detected and sites of pollution are known. Waters lacking in fertility show up and fertilization may be advocated. Soil erosion and related factors that may limit fish production are demonstrated and may be corrected. Situations of over-fishing and under-fishing are ascertained. Habits of fishes are disclosed of which advantage may be taken to increase yield. These and many other features make of fishery surveys in a broad sense a very important tool for the fishery management.

The conduct of fishery surveys involve techniques and specific kinds of information from diverse fields, such as, hydrology, surveying, geology, chemistry and limnology. It also calls for special knowledge on the part of the investigator in the aquatic phases of botany, entomology and invertebrate zoology. In addition it demands that one be thoroughly conversant with principles and techniques peculiar to fish management and that he know the fishes of his region well. Extensive knowledge in all of these fields is difficult for any one individual to attain. Most workers will often therefore find it necessary to seek the help of specialists. To be a good man in the field, one must have some basic training in each of the subjects listed above, and, before he can lead a field party, he should have considerable practical experience. He should have sufficient training in each of the subjects so that he is fully cognizant of the limitations of his knowledge and can recognize when things are going right and when they are going wrong.

Some steps of importance in the preparation and operation of a survey are the following:

1. Budget
2. Personnel, temporary and permanent
3. Organization and equipment of field parties
4. Planning, coordination, and integration of field program
5. Provision of base maps
6. Assembling of basic data already available from literature and from interviews and conferences
7. Initiation of intensive investigations where needed on problem lakes and streams, on pollution areas, or on productivity, migrations and other key life history features of important fish species.
8. Compilation of survey results and relation of findings to the definite objectives sought.
9. Publication of results in technical papers and popular reports.
10. Organization of supplemental research in a long-range program.

Impoundment Surveys (including lakes and ponds)

The procedures for survey should include, mapping of the site area, chemical analysis of water, studying plankton and bottom food samples, analysis of the aquatic vegetation, examination of existing parasites. Fish samples from the water are to be collected along with their scale samples to determine their age and growth. Previous stocking records of the water area, if any, may be collected along with the fishing history, with the objective to determine what may be done to improve fishing and the management of the body of water for higher fish production.

The study of a lake or pond is more difficult in many ways than that of a stream, since data cannot be collected as readily and since interpretation is likewise often more tedious. Certain stations are to be selected on each body of water at each of which the following data should be collected;

1. Temperature of the air
2. Temperature of the surface water
3. Temperature of water at various depths

4. Color and turbidity of the water
5. Character of the bottom
6. Water chemistry, depths and aquatic plants
7. Where necessary, counts of invertebrates from dredged samples
8. Qualitative and quantitative analysis of plankton hauls.

It may be assured to indicate the station for each operation by the proper symbol on the hydrographic map.

The number and location of the stations will depend upon the size and character of the body of water in question. One or two stations will be sufficient for very small ones, whereas, very large ones may require ten or more. The aim of the inventory should be the realization of the best possible cross section of the physical, chemical and biological conditions of the water area commensurate with the time available. Many important and possible phases of lake survey are often reduced to a minimum or omitted entirely because of lack of time or because of obscure significance; examples are analysis of plankton and bottom fauna. In addition to those made at complete, regular stations, observations should be taken at various points along the shores in order to determine the type of bottom, areas of vegetation, relative abundance of shore foods, presence of fish, potential or actual spawning grounds *etc.* It has been the policy in surveys to place the first station rather near the outlet, and if a large tributary stream enters, another station several hundred meters from its point of entrance. Since a complete picture of aquatic conditions cannot be obtained except through continuous study, a preliminary survey of this nature can do little more than indicate the general character of a particular body of water.

The name of the body of water as given on the base map in use should be employed.

Mapping

Sometimes contour maps of the lake to be surveyed will be already available. When no previous hydrographic maps are on hand, a plane table survey should be made before attempting the other analysis indicated for in the inventory. Two methods of plane table mapping are suggested. Sometimes one, sometimes the other is most desirable. A combination of the two often works out to advantage.

Setting up a Scale

The size limits of a map may be set for convenience as follows ; Minimum length of body of water on map 12 inches; maximum for sheet of paper, 24 by 36 inches. A convenient scale should be adopted for each water area. The scale in km. is to be included, also the legend, date, name of body of water, district and state on each finished map, as well as the source of the outline and soundings, if not made by the investigator should be given on each finished map.

Base Line Method

For establishment a base line, choose two points A and B as widely separated

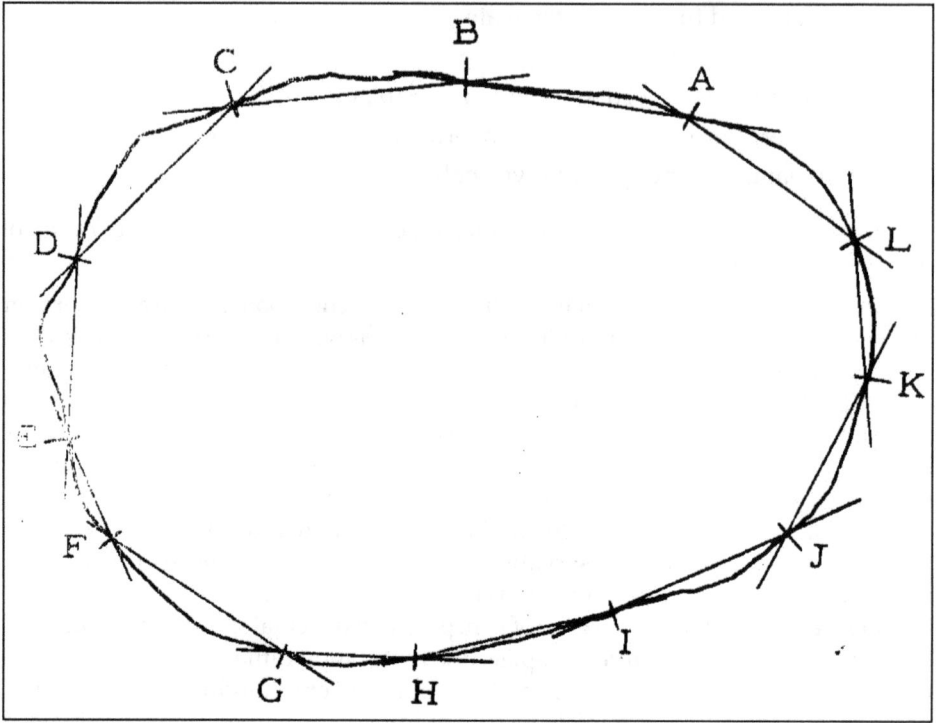

Shore Stations Used in Mapping a Lake of Regular Outline.

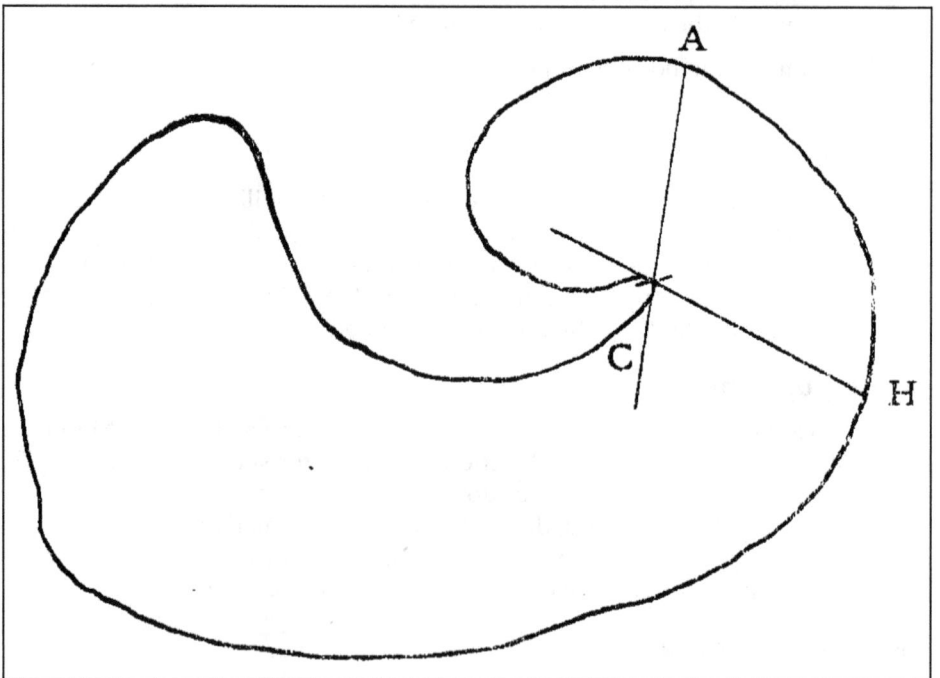

Emendation of Procedure to Compensate for Major Shore Irregularity.

as is convenient (100 to 300 meters), near the lake shore, from which most of the lake can be seen. Measure straight line distance between these points, preferably by means of a steel tape. Locate this line in such a position in relation to the lake that all angles of bearing lines will be as near right angles to the base line as possible. If the angles get small, establish other base lines and tie them in to the first base line or other reference points by back-sighting.

The plane table is set up directly over one end of the base line, the tripod legs spread well apart and firmly planted in the ground with the table approximately waist high. The table is then carefully leveled with a spirit level and oriented so that one side conforms to magnetic north. This procedure is to be followed each time the table is set up anew, including times when it is removed from place to place on a given body of water.

A convenient point on the paper fastened to the table is marked to represent the locus over which the table is set and is labeled point A. Using point A as a center, sight with the alidade to the other end of the base line and with a sharp, hard pencil draw a fine line toward point B. With the known length of the distance between points A and B, determine a suitable scale and mark off on the line AB line the distance, according to scale, between A and B; label the second point B.

From the point A now sight around the shore line of the lake from left to right bringing the alidade to bear on each major point and drawing a line from A toward each point. Label these lines 1, 2, 3, *etc.*, in order, and record to what each number refers (example : No. 1, dead tree on point; No. 2, large rock on shore *etc.*). Complete the circuit of the lake in this manner and then move the plane table to point B. Reorient and level the table at point B and verify by taking a back-sight from point B to point A making sure that the line between A and B corresponds to the edge of the alidade when point A is viewed from B.

From point B now sight around the lake as before to the various objects which were viewed from point A, drawing a line toward each from point B and numbering each line to correspond with the lines from point A. The point at which the lines intersect when extended is the location of the points desired.

If the lake is small and the shore line entirely visible from the two points this will be enough stations. If part of the lake is obscured, it will be necessary to proceed to some point already located by the intersection of two lines, to reorient the plane table correspondingly, and then to set up a base line and map the part of the lake not visible from the other station.

After all of the various points have been located by intersection, the details of the shore line may be filled in by careful free-hand sketching while moving from point to point.

Soundings are located on this map as they are made and the location and depth of each recorded. Analyze and record distribution of various bottom types, and vegetation, while making soundings.

The location of points where soundings are made can be done by using a base one or more of the known stations on the shoreline. While one man draws in the points by sighting with the alidade another stays on shore and keeps the boat in line between two other known points. Keep the soundings numbered in sequence so that they can be located on the map. This will have to be done on days when there is a minimum amount of wave action.

Traverse Method

The plane table should be set level and carefully oriented as to north and south direction (the weighted end of compass needle points south on tables with built-in compasses). The table is not level if the needle extends above or below the fixed points at other end.

Set plane table over station A, orient and level board. Place a small x or a common pin at a point on the paper in a position so that as the map progresses it will not run off the board. Place the O mark of alidade against pin using it (the pin) as a fulcrum. Swing the alidade until station B is sighted. Draw a line, measure the distance between stations and plot on the board according to the chosen scale. Then set up over station B and draw in the shore line between A and B, being careful to check by measurement any great deviations or irregularities from the straight line between stations.

It is well to take a back-sight on the previous station as a check on the accuracy. This is done by placing the alidade on the line drawn between stations and orienting and leveling the plane table, then sighting the previous station. The line of sight should bisect the former station.

On the finishing of the traverse, with the table set on last station (L) and properly oriented and with the alidade placed so its edge bisects both last and starting station (L and A), the line of sight should bisect station A and the measured distance should correspond with the plotted distance. This is a final check as to the accuracy of the map.

Suggestions while Mapping

☆ Take care of the alidade; rough handling can throw it out of adjustment.

☆ Be sure to check your closing error when finishing the traverse.

☆ Check often by back-sighting.

☆ In swinging the alidade to the line of sight between stations be extremely careful not to throw the plane table out of orientation.

☆ Remember a very slight error in the compass adjustment makes a large angular error on the map.

☆ A common or insect pin makes a fine tool to mark the stations and permits the alidade to be adjusted to the station point quickly.

☆ Use a hard pencil well sharpened. A coarse line makes accurate work impossible.

☆ Have a good system of keeping track of points and lines and intersections, name them if possible so that those plotted at one station may be easily identified when the table has been moved to some other station.

☆ Keep the bottom of the alidade clean.

☆ At the first set-up in a survey, draw a north and south line by means of a compass.

☆ One of the most important precautions in all plane table work is to take check sights frequently.

☆ Common sources of error are; position of plane table; plane table not level; faulty sighting; faulty adjustment; poor or confused lines; incorrectly plotted distances; contraction and expansion of paper.

☆ Be careful in the choice of scale.

☆ Be sure to lock compass needle while taking table from one station to another.

☆ Be sure to use standard legend.

Symbols to be Used in Mapping

In general it is well to use standard symbols for designating various features on maps. Sometimes, however, departures are made to fit particular circumstances. The symbols and practices are in common use showing in Figure on next Page.

Classification of Bottom Types

An outline and definition of common homogeneous bottom types are given below. Symbols for each category usable for mapping and for recording in field notes are also given.

Organic

Detritus - Undecomposed woody or herbaceous debris = D

Peat - Brownish to greenish, partially decomposed plant remains = P

Fibrous peat - Composed of coarser, herbaceous material, parts of plant readily distinguishable == fP

Pulpy peat - Uniform fine texture, parts of plants not distinguishable = pP

Muck - Black, completely decomposed organic material, ordinarily found in flooded areas or at mouths of streams = Mk

Inorganic

Bed rock - Rock strata *in situ* = BR

Boulders - Rocks over twelve inches = Bo

Rubble - R

Coarse rubble - Rocks six to twelve inches = cR

Fine rubble - Rocks three to six inches = fR

Gravel - Gr

Coarse gravel - one to three inches = cR

Vegetation types

Floating = ⊤   Emergent = ⊥   Submergent = —

Bottom types

Pulpy peat = orange, Fibrous peat = Light blue, Muck = Brown,

Sand = Yellow, Clay = Gray, Marl = Green

Gravel = ⟨gravel symbol⟩   Spawning beds = ⟨dotted symbol⟩   Brush shelter = ⟨symbol⟩

Spawning boxes = □   Snags, deadheads, etc. = ⟨symbol⟩   Trash = X

Outline and contours

Shoreline = ⟨line symbol⟩   Contours = ⟨3' contour symbol⟩

Shore features

Trail = ⟨dashed symbol⟩   Road = ⟨double line⟩   Cottage = ⟨black square⟩

Steep slope = ⟨symbol⟩   Encroaching shore = ⟨symbol⟩   Brush shore = ⟨symbol⟩

Marsh = ⟨symbol⟩   Cultivated land = ⟨Ⓒ⟩   Pasture or cleared land = ⟨Ⓟ⟩

Semi-wooded = ⟨PW symbol⟩   Tree = ⟨symbol⟩   Permanent inlet = ⟨symbol⟩

Intermittent inlet = ⟨symbol⟩   Outlet = ⟨symbol⟩   Spring = ⟨symbol⟩

Dock = ⟨symbol⟩

Stations

Vegetation with station number = ⟨①⟩   Plankton = ⟨symbol⟩

Temperature and chemical analysis = ⟨△ symbol⟩   temperature, chemical and plankton = ⟨△ symbol⟩

Bottom sample = ⟨[3] symbol⟩   Fish sample = ⟨fish symbol⟩

Symbols to be Used in Mapping.

Fine gravel - One-eighth to one inch = fGr

Sand - Particles smaller than fine gravel = Sd

Silt - Less compact than clay, very slight grittiness = St

Clay - Compact, sticky = C

Marl - M

Concretion marl = cM

Shell marl - Abundance of shells or shell fragments = sM

Amorphous marl - Uniform, relatively pure marl clouding the water when disturbed = aM

Classification of Aquatic Plant Types

There follow a few examples of the three ordinary categories of aquatic plant types; mapping symbols previously given.

Floating

Duckweeds (*Lemna, Spirodela, Wolffia*)

Water shield (*Brasenia*)

White water lilies (*Nymphaea*)

Spatterdock or yellow pond lilies (*Nuphar*)

Emergent (Leaves mostly emerging from water)

Arrowheads (*Sagittaria*)

Water (arrow) arum (*Peltandra*)

Pickerel weed (*Pontederia*)

Bulrushes (*Scirpus*)

Bur-reeds (*Sparganium*)

Cattails (*Typha*)

Spatterdock or yellow pond lilies (*Nuphar*)

Submerged (Leaves mostly submerged, a few floating leaves present in some)

Pondweeds (*Potamogeton*)

Wild celery or eel grass (*Vallisneria*)

Coontail or hornwort (*Ceratophyllum*)

Water buttercup (*Ranunculus*)

Water milfoil (*Myriophyllum*)

Waterweed (*Anacharis*)

Bushy pondweed (*Najas*)

Completing the Map

On tracing cloth or paper, a complete map may be made. If the scale chosen in the field was not suitable, this time it may be changed to give a map of the dimensions previously designated (minimum length of body of water, 30 cm; maximum size of sheet, 60 by 90 cm). Ink the map carefully and include complete legend : name of water area, general and exact location, scale, date of mapping and authority, true and magnetic north, key to symbols (and colors) used, and other pertinent data (study stations *etc.*).

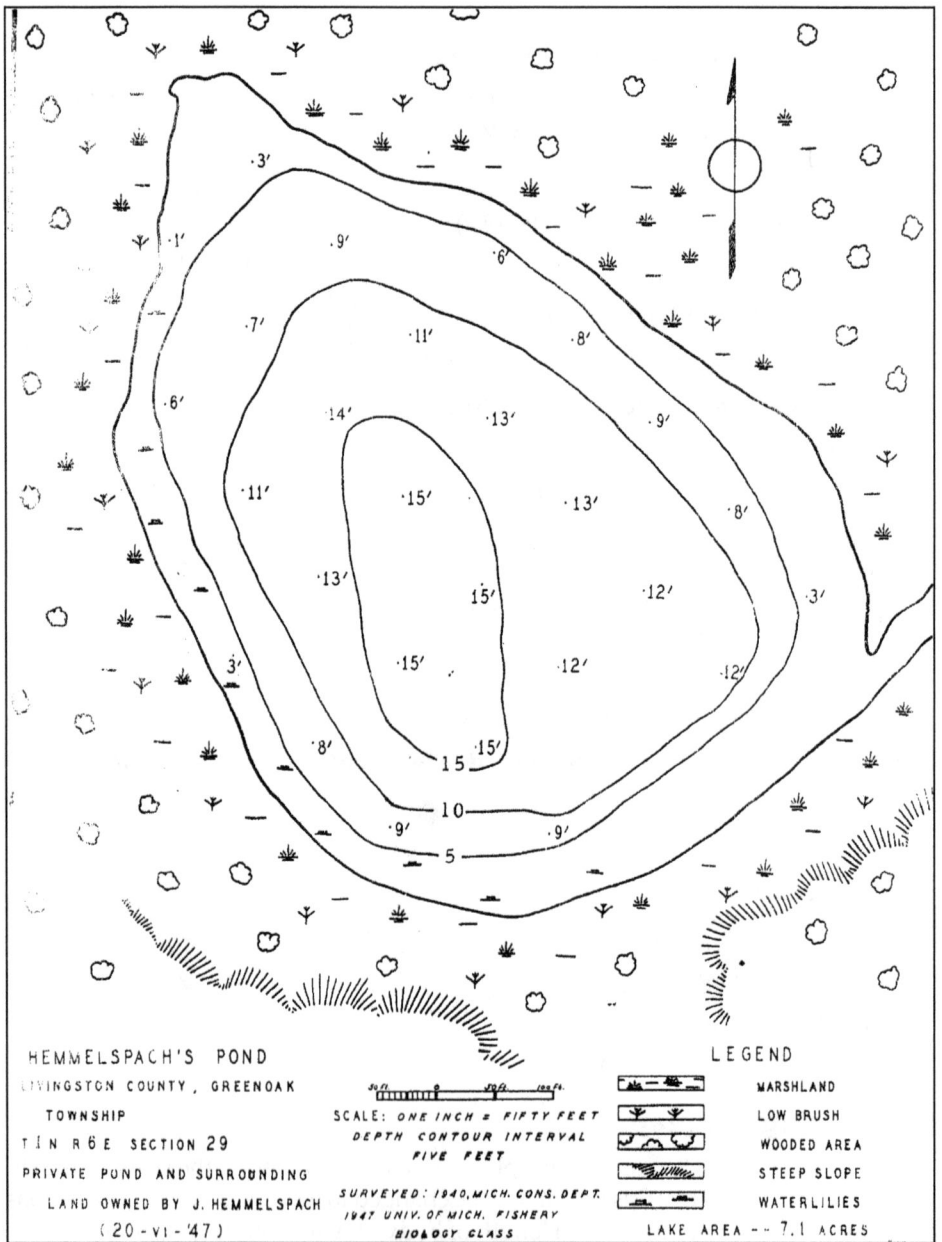

Sample Lake Survey Map.

Significance of Map

An accurate hydrographic map of a lake, pond or impoundment is not only of importance in the actual conducting of a fishery survey, but it is the basic to the formulation of the management plan following the survey. Some management procedures that require a map are : (a) fish poisoning- volume of water must be known in order to compute amount of poison to use and a knowledge of depths is required to enable proper distribution of the poison; (b) stocking – size, depths and adequacy of spawning grounds (bottom types) are important in determining what species may survive when introduced; (c) regulation of water levels- the desirability and feasibility of permanently raising or lowering water levels may be disclosed partly by the map; in reservoirs, the effects of raising and lowering the water level may be determined in part from a good map; (d) structural improvements – locations for improvement structures (brush shelters, *etc.*) and numbers of such to use may be judged from a map; (e) fertilization - if fertilizers are to be added to the water its volume must be known in order to calculate amounts to be used.

Chapter 16

Chemical Water Analysis

Chemical water analysis in routine fishery surveys are ordinarily restricted to the tests for dissolved oxygen, carbon dioxide, phenolphthalein alkalinity, methyl orange alkalinity and pH. Certain particular investigations may call for other important determinations such as nitrates, phosphates, *etc.*

In lakes and ponds, the exact number of chemical stations to be used is left to the discretion of the investigator. On very small lakes a surface and bottom sample at one station might be entirely sufficient. It is urgent to determine in all cases whether or not stagnation occurs in lakes. Analysis on larger lakes will therefore have to cover all major depressions. It is also urgent that analysis be made in the mouths of the larger inlet streams. The chemistry record sheet is adapted for recording chemical data of more than one series at a station. Impounded waters are treated in the same manner as natural lakes and ponds. In streams, one complete series of analysis should be made for each station.

Reagents

Make sure that you have a complete set and an adequate extra supply of reagents before entering the field. All reagents can be taken as ready-made (except for sodium thiosulphate whenever the survey is of such a nature that it will keep away from base for some time). Reagents in the following list concerned with determination of dissolved oxygen are for the Rideal-Stewart modification of the Winkler method.

Preparation of Sodium Thiosulphate Solutions

1. Ampules of sodium thiosulphate may be carried into the field. Each ampule should be made up to contain 3.1025 g of sodium thiosulphate which is enough for 500 ml of 0.025 N solution of sodium thiosulphate.

Reagent and Specification Amount

Concentrated sulfuric acid, 1.83-1.84 Sp. Gr.	One 250 ml bottle, glass stoppered
Potassium permanganate soln. (6.32 g/liter)	500 ml in brown bottle
Oxalate soln.(2 g Pot. Oxalate/100 ml distilled water)	500 ml
400 g Manganous sulfate in 1 liter distilled water	500 ml
700 g KOH and 150 g KI per liter	1500 ml
Sodium thiosulphate ampules (3.1025 g/ampule)	6 ampules
Pot. Dichromate soln. (1.225 g per liter)	250 ml
Soluble starch	10 g
Chloroform	25 ml
Phenolphthalein indicator	50 ml
Methyl orange indicator (0.2 g in 400 ml dist.water	50 ml
N/44 NaOH	250 ml
N/50 Sulfuric acid	3, 1-liter bottles
Distilled water	3, 1-liter bottles
Requisite pH indicator soln.	2, 25 ml bottles each

2. Remove label from ampule. Wash thoroughly to remove all traces of glue.

3. Insert ampule into 1 liter bottle and shake well until ampule breaks. Use clean, brown glass bottle. Be careful that the ampule does not break the bottle during the shaking.

4. Add 500 ml distilled water; mix thoroughly.

5. Should be provided with 2 brown glass, one-liter bottles for sodium thiosulphate solutions. While one sample is being used another should be made up and allowed to stand. This solution should be prepared about one week before it is to be used.

6. The solution from a given preparation may be used for two weeks. Standardizations are necessary about every 5 days or at any time when results appear ambiguous.

7. Be sure to have sufficient ampules on hand.

Standardization of 0.025 N Sodium Thiosulphate

The strength of the sodium thiosulphate solution must not be assumed to be 0.025 N, after it has been prepared. It may not be of the desired strength due to minor variations in preparation or differences in aging. It should be standardized as follows and a correction factor obtained for use in adjusting values gained in dissolved oxygen analysis.

1. A standard solution of potassium dichromate should be carried in the field.

2. Draw 10 ml of 0.025 N dichromate solution into a 100 ml beaker from a burette.

3. Immerse the beaker in an ice bath (at least in very cold water).

4. Add slowly (drop by drop) 1.0 ml. of alkaline KI. The same alkaline KI as used in the oxygen determination. Rotate the beaker during the addition of the alkaline KI.

5. Add slowly (drop by drop) 1.0 ml. of concentrated sulphuric acid. Rotate the beaker during the addition of the sulphuric acid.

6. Cautions

   i) Reagents should not be added too rapidly.

   ii) Mixture must remain cool.

   iii) Operator should not detect the odor of iodine.

   iv) Failure to heed the above cautions will introduce a serious error.

7. Titrate the above mixture with the sodium thiosulphate to be standardized.

   i) When the solution becomes pale yellow add a few ml of clear starch solution.

   ii) Titrate to the end point and record amount of sodium thiosulphate used. The end point is not as definite as it is for the oxygen determination. The sample retains a green color even after the end point is reached (experience will bring about an acciurate judgement of the end point). Repeat at least three times and average results for use as a correction factor in dissolved oxygen determinations.

Results

i) If in the standardization exactly 10 ml. of the sodium thiosulphate was used then the thiosulphate solution is exactly 0.025 N.

   If the sodium thiosulphate is exactly 0.025 N then the titration value in the oxygen determination is equal to oxygen in parts per million (ppm).

ii) If the amount used in the standardization varies on either side of 10 ml. a correction factor must be made.

Variation I

(a) If 12 ml were used in the titration the solution is less than 0.025 N. Therefore the titration value in the oxygen determination will be too high. The correction factor, therefore, should be less than 1.

(b) 10 divided by 12 equals 0.833. 0.833 is now the factor in this case. Multiply the titration value in the oxygen determinations by the factor in order to get the oxygen in ppm. Example : if the titration value in an oxygen determination is 12 ml. multiply this by 0.833 to get a corrected value of 10.0 ppm for oxygen.

Variation II

(a) If 8 ml. were used in the standardization, the sodium sulphate is more concentrated than 0.025 N. Therefore the titration value in the oxygen determination will be too low. The factor, therefore, should be more than 1.

(b) 10 divided by 8 equals 1.25. 1.25 being the factor in this case. Multiply the titration value in the oxygen determination by the factor in order to get the oxygen in ppm. If the titration value in the oxygen determination is 8 ml., multiply this by 1.25 which equals 10.0 ppm, of oxygen for the corrected value.

Dissolved Oxygen

There are several methods for determining dissolved oxygen content of water (Welch, 1948; Ellis, Westfall, and Ellis, 1946; Standard Methods for the Examination of Water and Sewage). The Rideal-Stewart (permanganate) modification of the Winkler method is among those reasonably well suited for use in ordinary, inland freshwaters. This modified method has been widely used in fishery surveys. The procedure is as follows. But before starting the oxygen sequence, however, it has to be made certain, that thiosulphate solution has been standardized as described earlier.

1. Collect sample into 250 ml. glass-stoppered bottle and record station, depth, water temperature at the given depth *etc*. The bottle (Juday bottle) should be equipped with a glass or rubber delivery tube and the delivery tube should be inserted deep into 250 ml. bottle to avoid contact with air. 250 ml. bottle should be overflowed two times its normal capacity (20 seconds after the bottle is filled). Glass stopper should be carefully inserted to avoid air-bubbles in the neck of the bottle. The bottle number should be recorded on record. Reagents should be added as soon as possible to avoid changes in oxygen content.

2. 0.7 ml. of concentrated sulphuric and 1.0 ml. of potassium permanganate to be added using separate pipettes for these and all other reagents. Pipettes must be dipped below the surface of the sample.

3. Restoppered and mixed. Allowed to stand for 20 minutes. If the purple color still persists at the end of 20 minutes, step 4 may be proceeded. If this color does not persist, 0.1 ml. more potassium permanganate may be added and allowed to stand.

4. If the color lasts for 20 minutes, 1.0 ml. of potassium oxalate may be added, restoppered and mixed.

5. After the sample is perfectly clear, 1.0 ml. of manganous sulphate may be added along with 3.0 ml. of alkaline potassium iodide, re-stoppered and mixed.

6. Allow precipitate to settle.

7. 1.0 ml. of concentrated sulphuric acid may be added and mixed. If the sample is securely stoppered it may stand several hours without a change in the oxygen content, but it is better to titrate it as soon as possible.

8. Titration.

9. Excess water over 200 ml from sample bottle.

10. Titrate remaining 20 ml of sample with 0.025 N sodium thiosulphate. The bottle may be rotated during addition of sodium thiosulphate until the sample becomes a pale yellow.

11. When the sample becomes a pale yellow, 1.0 ml of starch solution may be added; the resulting mixture will ordinarily be dark blue.

12. Continue titration carefully until blue color disappears. Record the amount of sodium thiosulphate used during titration.

13. Ml. of 0.025 N sodium thiosulphate equals dissolved oxygen in ppm. The correction factor may be used, obtained from standardization of thiosulphate to obtain corrected value for dissolved oxygen when factor does not equal 1.0. The results may be recorded.

On completion of dissolved oxygen analysis, it is to be considered whether or not the results obtained are reasonable before proceeding with other steps in the survey. The duplicate sample should run for the same results. Ordinarily, dissolved oxygen values will not exceed those for saturation at any given temperature. This may be checked from the nomogram, obtaining saturation values of dissolved oxygen. To see if the sample is within the limits of saturation as indicated there. If the results indicate super-saturation (a rare condition), the following simple test may be employed. The analysis may be repeated, but immediately after placing the water sample in the sample bottle, the bottle may be agitated violently. If the amount of dissolved oxygen is reduced by this procedure, it is likely that a condition of super-saturation was originally encountered. If not, solutions and routine have to be checked to find the cause of error.

Significance of a Knowledge of Dissolved Oxygen Value

Amount of dissolved oxygen present at various times is determined in fishery surveys of standing waters because of its importance in indicating environmental suitability, mostly for cold water fishes. For example, in order that lakes which stratify thermally in the summer may support trout, there must be sufficient oxygen in the stagnant water below the thermocline. If such water (of suitable temperature and containing 4 to 5 ppm of oxygen during the summer stagnation period) does not exist in a particular body, there would be little point in trying to introduce or encourage trout in it.

Free Carbon Dioxide

1. 100 ml. of sample may be drawn into a Nessler tube. The water should be flowed along the side of the Nessler tube to avoid any unnecessary agitation. Agitation may change the amount of dissolved carbon dioxide. The sample may be taken from the Juday bottle directly after collection of oxygen sample.

2. Ten drops of phenolphthalein may be added as an indicator.

3. If the sample turns pink, 0.0 may be recorded.

4. If the sample remains clear, titrate with N/44 sodium hydroxide from a burette, until a weak pink (pink color should remain at least for 30 seconds). The number of ml. of sodium hydroxide used in titrarion may be recorded. Mixing can be accomplished by placing hand over open end of Nessler tube and inverting the tube.

5. Results : Ten times number of ml. N/44 sodium hydroxide used equals amount of free carbon dioxide in ppm.

Significance of a Knowledge of Free Carbon Dioxide

Determination of the amount of free carbon dioxide in water is important in fish management because it is the best single criterion of environmental suitability for fishes. Furthermore, high concentrations of free carbon dioxide which are in themselves toxic to fish are usually accompanied by low values for dissolved oxygen. In general, free carbon dioxide in excess of 20 ppm, may be regarded as harmful to fishes although lower values may be equally harmful in waters of low oxygen content (less than 3 to 5 ppm).

pH

An acceptable field method for determining pH values of water, besides battery operated portable electronic pH meter is the Hellige glass comparator. To give best results, the indicator solutions and corresponding comparator discs should have a substantial overlap in series.

1. Ten ml of the sample may be drawn into each of two pH tubes. This analysis can be made on water from part of the sample used for oxygen and carbon dioxide.

2. Add the required amount of indicator (indicated on bottle) to one of the tubes. The other tube may be placed in the comparator without indicator to correct the natural color of the water.

3. Both the tubes are then placed in the comparator.

4. By revolving the disc corresponding to indicator solution used, the colors are matched.

5. **Results** : pH can be read directly from disc when colors match and can be recorded.

6. If possible, may be verified with indicator of overlapping range.

7. **Cautions** : It is to be made sure to keep indicator solutions away from heat, cold and light as far as possible. A separate pipette may be used for each indicator. Excess solutions in pipettes on ground may be avoided, not to return to bottle and not to take indicator up into bulb of pipette. Can be mixed by rotating test tube and not to contaminate contents by placing them in contact with skin of fingers. Pause to determine whether or not the pH values are within the limits of normal expectation, about 6.5 to 8.5 in ordinary waters. Now consider them in relation to the amount of free carbon dioxide found to be present in the same water. Ordinarily if pH values are less than 7.0 some free carbon dioxide should also be present

in the water sample; it is often present when pH values are between 7 and 8. If pH is greater than 8.2, carbon dioxide does not exist free in the same water. Later, when phenolphthalein alkalinity is determined, reconsider the for free carbon dioxide and pH to see if they are consistent with the following generalization. Phenolphthaleion alkalinity is not indicated in samples where free carbon dioxide is present. In other words if the sample is pink with phenolphthalein indicator, no free carbon dioxide is there, but instead there is an indication of the presence of carbonates, the amount of which can be determined by titration with 0.02 N sulphuric acid.

Significance of pH Values

Determinations of pH are mostly valuable for detecting pollution (possible source of pollution or other unusual explanation is sought when values are outside the 6.5 to 8.5 range). Most fishes tolerate both wide ranges and rapid changes of acidity and alkalinity as expressed on the pH scale.

Alkalinity

1. From the water remaining in a Juday bottle, 100 ml of the sample may be drawn into a 250 ml glass-stoppered bottle (bottle with a 100 ml. graduation mark).
2. Four drops of phenolphthalein indicator may be added to the sample.
3. If the sample remains clear, 0.0 ppm ph-th may be recorded.
4. If the sample becomes pink, titrate it with 0.02 N sulphuric acid, from a burette, until it becomes clear. The bottle should be rotated during the addition of the acid. The amount of acid used may be recorded.
5. To the same sample, 2-6 drops of methyl orange indicator is added.
6. If the sample turns orange 0.0 methyl orange may be recorded.
7. If the sample remains yellow rather turning orange, it may be titrated with 0.02 N sulphuric acid, from a burette, until it turns orange. The bottle should be rotated during the addition of the acid. The amount of acid used in the M.O. titration including the amount used in the ph-th titration. In order to identify the delicate end point of this reaction, it is helpful to set up a comparative sample in another flask which is not to be titrated but which can be used for color references.
8. Results : (a) 10 times the ml. of acid used in the ph-th titration equals ph-th alkalinity ppm as calcium carbonate.(b) 10 times the total ml. of acid used in the ph-th plus M.O titration equals M.O. alkalinity in ppm as calcium carbonate or total hardness

Significance of Total Alkalinity in Fish Management

Ordinary freshwaters may have as much as 350 ppm. Total alkalinity or total hardness expressed as calcium carbonate, although most values will be between 45 and 200 ppm. (Ellis, Westfall and Ellis, 1946). The authors cited and generalized that bicarbonates and carbonates in these quantities have little effect on fishes. However, knowledge of the values enables the classification of waters into very soft water (0-5 ppm), soft water (5-10 ppm), medium water (10-20 ppm), medium hard water (20-30 ppm) and hard water (more than 30 ppm), with the general knowledge that soft waters are less biologically productive than hard waters.

Chapter 17

Biological Investigations

Plankton Samples

In ordinary routine surveys, plankton samples may be omitted since so little can be learned from them. However, plankton swarms or absence of plankton should be investigated.

Care of Plankton Net

The plankton net is made of exceedingly fine (No. 20) silk mesh. Care must be taken to maintain it in a condition of maximum efficiency. Any holes may be repaired by touching a very small drop of Duco household cement to each break. Allow to dry 15 minutes before using.

Before collection of plankton at each station detach bucket and carefully wash net by drawing it through water. Similarly the bucket should be washed taking care not to loose the stopper pin which is loose in the bottom of the bucket.

Collect plankton on opposite side of the boat from that on which net is washed. On completion of sample collection, the net and bucket may be washed again. Each night the net should be shaked and dried thoroughly, replacing net and bucket in separate pockets of the bag or other container which is provided.

Collection of Plankton Samples

1. Anchor the boat at the station.
2. Have plankton analysis form handy for recording data.
3. Take sounding to determine the depth.
4. Permit the boat to drift two or thee meters away from the region of any bottom disturbance.

5. After the net is washed on opposite side of the boat, use the following procedure for collecting the sample.

6. Subtract 2 meters from the depth at the station and lower the net to this point. Draw the net to the surface at the rate of approximately 30 cm (1 foot) per second. Be certain that the bucket does not touch the bottom as silt will ruin volume determinations. In lakes shallower than 2 meters a measured horizontal haul may be taken.

After the haul is made permit all water to drain into the bucket with about 10 ml. of water remaining in the bucket.

Place centrifuge tube under the bucket and drain water with a slight rotary motion into it. Rinse bucket by permitting water to come in through sides (not over top). Again drain remaining 5 ml. of water into tube. Add ½ ml. formalin to each of 10 ml.of plankton and water.

Place in tube at once and label it with data giving name of the lake, station number, length of haul, date and collector's name. Cork the tube carefully and replace in case. Allow each plankton sample to settle for approximately 24 hours; then read the volume to the nearest tenth of a ml. and record. Also record the number of ml. per cubic meter which should be determined by the following formula :

$$\frac{\text{Volume in ml. of plankton}}{0.00985 \text{ sq. m. (area of net mouth, 11.2 cm.dia)} \times \dfrac{\text{length of haul}}{3.28}} = \text{ml per cubic meter}$$

Substitute the volume of plankton and length of haul in feet and determine the result.

Bottom Food Samples

The number of Ekman or other dredge samples to be taken in each body of water will depend on its size, suitability of bottom sampling, and the time available for the care of samples.All identifications and measurements should be made on the field unless other arrangements are in order. The total volume (by water displacement in a centrifuge tube) is recorded as well as the number of each kind of organism present and weight if obtainable. The sample should be made as representative as possible and a complete record should be made.

Shoal sampling will have to be limited to qualitative observations. Specimens not broken in the counting and measuring process may be preserved in 80 per cent ethyl alcohol for finer identification if this is desired or necessary in detailed surveys.

In normal practice, only enough samples need be taken to give an idea of kinds and relative abundance of organisms present. If there is evidence of stagnating or decline in the fish population, detailed bottom food analysis throughout the seasons may be necessary.

Aquatic Vegetation Survey

Because of the difficulties involved in the classification of aquatic plants it may be necessary to make extensive collections. Both a plant hook and a rake should be available for use.

One label should be filled out for each different plant collected at each station. The first line should have the collection number followed by symbol for abundance (S = Sparse; M = Medium: D = Dense). Following this should be given the area or extent of the coverage, namely, ¼ acre or a strip or a strip of 50 m. wide and 100 m. length *etc.* The number of station corresponding to the numbered symbol placed on the map may be recorded.

Approximate depth of the water at the point where the specimen was collected may be given with the range of depth for the species in question for the particular vegetation station. Under the bottom should be included the appropriate term like clay, pulpy peat, fibrous peat, sand, gravel, muck, marl *etc.*

Vegetation stations should be numerous enough to give a good cross section of the plants in the lake or stream. The exact number will have to be determined according to the best judgment of survey parties. Observations and corresponding station numbers on the aquatic vegetation may be recorded.

Plant collections that are meant to serve as records of the survey, enabling doubtful identifications to be verified whenever necessary in the future, must conform to certain standard requirements.

Since herbarium specimens are preserved on paper sheets 28.75 by 41.25 cm. in size, the amount of plant material of any species collected from any given locality should be sufficient to approximately fill at least one herbarium sheet without undue overlapping of parts. If the plant species seems to be rare and is in excellent condition (with flower or fruit), a few extra specimens (full herbarium sheets) should be collected. Only one sheet of each species from a given station is necessary for record purposes, unless the species shows much variation, in which case a series of specimens illustrating each marked variation should be secured.

The specimens selected should in general consist of complete plants so that all morphological parts are represented. When a complete plant is too large to come within the size limitations of a herbarium sheet even by folding or bending, only representative portions of the plant need be collected, or the plant may be divided into suitably sized consecutive parts which are considered as serial portions of the collection. The latter are to be labeled in order "a", "b", "c", *etc.* In the case of water lilies, for example, a representative set would consist of two leaves (or portions thereof) so that the upper and lower surface can be seen when they are correctly mounted, two or more flowers, a fruit (split open), and a split portion of the rhizome. In general, aquatics can be collected as complete plants. When they are longer than 37.5 cm, they are best bent into long N's or M.s and should not cover an area larger than 25 by 37.5 cm. Most pond weed collections should have mature fruit rather than flowers, if, as is often true, both cannot be obtained. It is true of aquatics in general that fruiting specimens are easier to identify than flowering ones, although both should be secured when possible. Moreover, it is well, whether collecting a single sheet or several sheets of small species, to put in several extra clusters of fruits or of flowers, whichever may be available. These add greatly to the permanent reference value of the material, since they permit it to be studied repeatedly without being destroyed.

Fish Samples

Though it is urgent to secure a complete cross section of the fish population, this must be done over a reasonable period. The methods used will be determined by the nature of the water to be studied. Often night seining is desirable and will yield many more fish (both numbers and kinds) for a given amount of seining effort than seining in the daytime. In lakes and impounded waters, seining should be supplemented by gill-net sets, by the use of fyke or trap nets and by examinations of angler's catches. Graded "experimental" gill-nets ensure a broader catch both as to sizes and kinds. Fish collecting stations should be indicated on the map by the proper numbered symbol and should be accompanied by complete records on a fish collection details.

Seines should include at least one 25-foot seine 6 feet deep with a trailing bag (bag-seine) of 6 mm square mesh. This is the best all-purpose net for shore seining in inland waters.

Certain identifications may be made in the field, but only for species readily recognized. Where any doubt arises, specimens should be preserved for later identification. Specimens of identifiable species need not be saved, but measurements and counts should be made on those discarded. When large series of those not identified are collected, only a good representative of sample, 25-50 specimens, should be saved. This information should appear on the fish collection record.

All length measurements should be consistent and accurate and of a kind using conventional, accepted criteria. They should be in cm. and tenths of cms, except for smaller fish which may be measured in mm.

All weights should be made on accurate balance in kgs. and hundredths thereof.

Fish collections should be properly labeled with a serial number corresponding to that placed on the fish collection blank. The label should also have the name of the water area and date. For labels, only a stout variety, resistant to the preservative may be used. For writing, soft lead pencil may be used. When collections consist of small fish, these with the completed label may be tied firmly but loosely in a piece of cheese cloth and saved together in a large container. Larger specimens should be individually labeled with plastic tags. These numbers should be recorded on the fish collection card. All fishes over 12 cm. long should be slit two cm. or so on the right side to insure proper preservation.

All fish collections should be preserved in ten percent formalin. Two liter jars and 50 liter cans are most useful for these collections. All containers should have a label securely tied to the outside as well as one inside. Labels on large containers should have a list of the collection numbers as well as the water from which these collections were made.

Scale Samples

Substantial series of scale samples should be secured for all important game and food fishes. Particular attention should be given to getting good representation in the largest and smallest sizes.

Previous Stocking Record and Fishing History

Records of previous stocking may be obtained which should include year and number and kinds of fish stocked. If available, a summary of creel census records may be obtained by interviewing fishermen fishing in the water area.

Survey Report

After completing the necessary laboratory studies to supplement field work, a running account of the survey using the following guide line. A brief statement as to the significance of each point of information for formulating management plans is to be emphasized. In doing so, a brief explanation of how each environmental factor studied might limit the fish production may be given.

Introduction

Location and accessibility

Nearby township

District

State

Other geographical features nearby

Other water areas

Drainage affinities

Accessibility to general public

Roads, Train *etc.*

Map and Survey

Map- a black-line map on tracing cloth suitable for printing

Source of map

Biological survey

Personnel

Time of survey

Purpose of investigation

History and Recreational Development

Amount of recreational development

Amount and nature of industrial development

History of Fishing

Past, present including creel census records

Boat liveries

Hotels and resorts

Importance as a public fishing water

During summer and winter

Importance other than as fishing water

Desirability of maintaining or altering present state of development

Physical Characters

General physical characters

Shape of basin

Area of surface

Length of shore line Compass direction of main axis

Depth, regular and irregular (describe depressions *etc.*)

Shoreline development (give figure, state its significance and how computed)

Dropoff, presence, absence, location

Shoal areas, widths around lake, percentage of surface area of lake made up by such

Bottom types and extent of types

Water color

Transparency, Secchi disc readings, sources of turbidity, if turbid

Geologic origin of the lake and present stage of evolution

Drainage of the lake

Extent of drainage basin of the lake itself in square Km.

Nature of drainage basin

Soil, ground cover, woods, marginal vegetation

Fluctuations of water level

Inlets, size, flow, dimensions, characteristics, points of entrance, differences among inlets

Use of lake fishes for spawning *etc.*

Outlets (as for inlets)

Dams in drainage system

Temperature and Chemical Characters

Significance of these

Water temperatures, surface, time

Air temperature at same time

Water temperature, bottom, at what depth

Thermocline

Oxygen, epilimnion. Thermocline, hypolimnion

Other chemical characters

Carbon dioxide

Methyl orange alkalinity – degree of softness or hardness

pH

Pollution

Biological Characters

Vegetation (composed of species list, relative abundance, and bathymetric distribution)

Fish foods

Plankton dominant forms, relative abundance

Bottom feeds, dominant forms, relative abundance, qualitatively and quantitatively

Food of the fish- compare with available forms

Fishes present

Game species – sizes, kinds, relative abundance

Food fishes – kinds, sizes, relative abundance

Forage fishes – kinds, sizes, relative abundance, significance in food of piscivorous fishes

Obnoxious fishes

Growth rate and condition of food fishes of the water area, comparison with regional averages and differences may be explained.

Adequacy of spawning facilities and natural propagation

Previous management result including stocking and fate of introduced species.

Management and Improvement

The improvement of standing waters has its major objective the production of maximum sustainable yield of preferred species. The method for achieving this objective is manipulation of the environment, the fish population, or both. Detection of the need for improvement lies in fishery surveys, often in very intensive ones.

Shelter

Shelter (other than aquatic plants) is seldom too abundant in natural waters. In general it favors survival by providing escapement and it tends to increase the substratum for fish food production. Shelter may also serve to concentrate fish of catchable size. Often there is need for more such cover and it can be provided by the introduction of brush shelters.

Aquatic Vegetation

Aquatic plants, especially seed-bearing ones, may be too abundant. As such, they may be filling the body of water too rapidly or favoring survival of fish to such extent that there is overpopulation and stunting of growth. Plants may be controlled chemically or mechanically. In reservoirs, water level fluctuation may be used to retard aquatic plant growth.

If aquatic vegetation is judged to be too sparse, introduction of seeds, tubers, or whole plants may be made or fertilization may be tried. It may be remembered that since fertilizers also increase plankton they are useful for controlling rooted aquatics by reducing light penetration through the water.

Spawning Facilities

Facilities for spawning may be so good that too many young are produced with resulting overpopulation. This may be controlled by nest or spawn destruction or by covering parts of spawning grounds with unfavorable materials. In impoundments, water levels may be altered during spawning seasons to decrease the spawning area (or to increase it if required)

If spawning facilities are inadequate, they may be augmented by placing suitable nesting materials in proper locations.

Water Level and Depth

Sometimes elevating the water level of a lake or pond will increase its fish producing capacity; low dams are useful for this purpose. At other times, lowering the water level, to increase the shoal area for fish food production, may be beneficial. In impoundments, it is often desirable to minimize fluctuations in water level, particularly during the spawning season so that spawning grounds are not alternately stranded and drowned.

Natural Food

Available food supply is basic in fish production. It is never too abundant. Plankton and insects may be increased by fertilization. Other foods of piscivorous fishes, such as, minnows, may be introduced successfully, particularly if care is taken to ensure adequate spawning and escapement facilities for them.

Erosion and Silting

Erosion and silting are two of the most important natural agencies tending to fill standing bodies of water and are of particular concern in impoundments. Generally, control is desirable. Very often this control involves planning and work in an entire watershed and may start with recommendations for contour plowing and soil stabilization in a farmer's field, kilometer away from the body of water in question. Eroding bluffs along shores may be stabilized by terracing, by planting soil-binding woody and herbaceous plants, or by the use of floating log booms to dissipate wave action.

Pollution

Pollution is seldom a problem in inland standing waters, except in impoundments. In small quantities, certain types of non-toxic organic pollutants increase fertility. However, if the body of water is also used for human consumption or for bathing, even the smallest sources of pathogen-bearing pollution (domestic sewage) should be arrested. Substances of known toxicity to fishes should forever be excluded from inland waters, particularly if they may produce toxic effects by accumulating (substances containing metallic ions, such as copper).

Seasonal Anaerobiosis

Complete oxygen depletion leading to mass fish mortality may occur during several seasons of the year; all such situations cannot be alleviated. Flash mortalities resulting from sudden, temporary spread of oxygen deficient waters at the time of spring or fall overturn in northern waters are apparently without remedy. Suffocation or poisoning of fishes resulting from sudden decomposition of large quantities of algae (plankton algae or filamentous algae) or protozoans is also difficult to combat. In small shallow bodies of water algal mats may be raked aside. Large algal blooms may also be chemically controlled, but the chemicals required are toxic to fish, if not in the quantity used in the single application, then in the accumulation resulting from repeated doses. The hazard of winter kill resulting from oxygen depletion may be reduced by mechanical aeration. In general these operations are costly and are applied where the existing fish population is in particularly good balance and especially prized.

Temperature

Shallow water bodies may become so warm that preferred food fishes will be in distress. Sometimes this condition may be improved by planting water or pond lilies which afford shade. In impoundments, stagnation may occur in the depths owing to the surfacing of warm waters. Such a condition may be overcome by drawing water from the bottom of the dam, rather than over the top of a spillway. Such drainage may be used to extend significantly the habitable waters for particular fish species, both within the impoundment and downstream from the dam retaining it.

Fish Species Present

Experimental work has re-emphasized the generally known fact that certain fish species in combination may produce more fishing of a desired quality rather than other species combinations. Some voids in fish production may be filled by introducing new kind of fish into a body of water, after environmental suitability has been established. In other instances, it may be necessary to remove the entire existing fish population and to start with a chosen, limited combination of forms. Impoundments may be drawn down to this, natural waters may be poisoned with rotenone. Poisoning may be partial or total.

Overpopulation

It is a normal characteristic of fish populations in smaller inland waters to be composed of more fish than could realize optimum growth on the basis of available food. It has furthermore long been known that a body of water supports only a certain number and weight of fish at one time; it obviously follows that if the fish are exceedingly numerous they will be of smaller average size than if they were less numerous. In the absence of adequate number of predators or fishermen, overpopulation leads to stunting, often severe. Stunting is identified by a substantial lag in rate of growth within a population as compared with an average growth rate for the region. Some remedial measures are; partial poisoning or other technical means of fish removal, discontinuation of stocking of the unbalanced species, introduction of predatory fish species, augmentation of fishing pressure and liberalization or removal of fishing regulations.

Predators and Competitors

In some situations, predator control may produce an increased yield of preferred species. This is an exception rather than the rule in inland waters, in spite of the popularity of such control with fishermen in general. Also, it should be recognized that predators may be beneficial (as in population balance) as well as harmful.

Competition has many ramified bearings on fish production. Works have shown that in spite of large quantity of coarse fishes present in a body of water, the total weight of predacious fishes at the same time was smaller than it would be were the quantity of coarse fishes less, Coarse fish removal may be indicated in some waters.

Diseases and Parasites

The roles of diseases and parasites in fish production in natural waters are essentially unknown. There are no known means for combating such conditions that could be recommended with complete confidence. As one preventive measures, fish with contagious diseases should not be stocked.

Basic Fertility

Ordinary agricultural fertilizers and manures have been used successfully for increasing fish production. However, there are no reliable formulas for determining exactly the kind and amount of fertilizer to be used for this purpose. Furthermore, the detrimental effects which fertilizers may have on aquatic plants demands caution. If fertilization seems desirable in a body of water, analysis should be made of available nutrients in the bottom and in the water and the kinds and amounts of fertilizer to be used planned from there. In general, repeated small doses would seem to be a better approach to fertilization than single large applications. Any fertilization program should be considered as experimental and should be supervised and followed strictly.

Fishing Pressure

Fishing pressure is, in a sense, an environmental factor which may be manipulated to increase both fish production and yield. The pressure may be too great or too small, in relation to the productive capacity of a body of water, to give maximum yield or annual crop. In many impoundments, it would appear that both the fishing pressure and the way it is regulated permit inadequate harvesting of the fish crop. This is not only wasteful but it may lead to the survival of less desirable species and contribute directly to overpopulation and stunting. In general, it is best to manipulate and regulate the fishing pressure individually for each body of water, changing the rules to obtain greatest possible sustained yield as the population changes. Often this will be impossible because of existing laws (which resist change) or public sentiment (which also resists change) and concessions will be necessary and expeditiously.

The following cautioning considerations have been suggested when environmental lake improvement is contemplated.

1. The lake has to be considered as an aquatic farm, the food fish as the desired crop. Common sense and planning such as is required in farm operation may be applied taking into view of seed, fertilizer, crop and harvest.

2. The principal factors which appear to be limiting fish production in the body of water is to be determined.

3. To meet the particular needs of the body of water has to be planed for improvement, and if feasible, detailed studies "before and after" may be planned and conducted to determine the effectiveness of what have been done.

4. Over-standardization of work has to be avoided by carrying out improvement projects, treating each lake as a separate management.

5. By building up those conditions which, because of their deficient development fish production are retarding has to be balanced for fish life in the environment.

6. Future needs have to be anticipated.

7. For the sake of efficiency, the work along one section of the shore line at a time may be concentrated.

8. When construction work is required, it may be done on dry lake bottom where feasible as in new impoundments.

9. Improvements may be made where maximum results may be expected.

10. Beneficial situations may not be destroyed while making improvements.

11. Material in or along lake when available may be used.

12. Siltation from surrounding lands may be retarded.

13. When desirable fluctuations in water level may be minimized, particularly during spawning season.

14. Building has to be done for permanent natural appearance.

15. Improvement practices have to be integrated with other recreational use of the waters.

Chapter 18

Creating New Fishing Waters

The creation of new fishing waters is an important development in fish management. Such waters are essentially of two kinds; (1) small, artificial reservoirs including farm fish ponds of multiple uses; and (2) ponds or other water areas in which a particular kind of fishing is seasonally provided, often close to centers of population and often dependent entirely on stocking. The latter type is extremely popular, becoming more and more numerous- several thousands are already in existence. Farm ponds are usually private and are built by farmers as a part of a soil erosion control program and as a source of water supply for livestock, with fish production as a secondary interest. Other small artificial reservoirs are publicly owned and in addition to fishing possibilities they have many other uses, such as bathing, swimming *etc.*

The basic requirements in the creation of any such small bodies of water are:

1. The soil must be of a kind that will hold water in the impoundment and not let it seep away.

2. The water supply- springs, surface run-off (area of drainage basin), tributaries- must be of sufficient magnitude to offset losses by evaporation, seepage *etc.* and by planned uses (livestock, irrigation *etc.*).

3. Associated erosion hazards must be minimized, usually by planting suitable vegetation, to retard filling by silt, and to protect dam from washing out and other banks from slipping as a result of wave action. Sometimes fencing will be required.

4. Some marsh and aquatic plants that compete least with, or favor most the uses to which the pond is to be put may be introduced.

5. Stocking with fish should be conservative and should be carried out with the thought in mind of the beneficent effects of short food chains on fish

growth. For this reason carps and catfishes have proven to be a good combination, particularly when no other fish are introduced with them.

6. If by any means possible, the dam should contain a well-built control structure of adequate size to pass flood waters. A means of draining the pond to control excessive plant growth or over-population by fish is desirable.

7. Opportunity for use of such ponds by ducks and other wildlife should be considered.

8. Fertilizer may be applied if the need is determined and if the probable effects may be predicted.

9. Best fishing is likely to ensue if the cropping by seining is heavy enough to prevent over-population with resultant stunting of fish.

10. Owners or builders of farm ponds should consult fish managers or other qualified persons freely to avoid costly blunders which may be made because of the ignorance of the experience of others. This is particularly pertinent when it comes to choosing the kinds, numbers, and sizes of fish to plant; suggestive stocking ratio does not always work well because of the individuality of waters.

Field Program

A farm fish pond or a small artificial reservoir may be visited and the questionnaire given below may be filled in. Information from the owner and builder may be secured. What consideration has been given to the principles previously described may be determined and the result of the application of or failure to employ these principles in the development of the project may be identified. Finally, using the references and other information, the best fish management proposals compatible with the other uses may be made of the pond.

Evaluation of a Farm Fish Pond or Small Reservoir

Location _____

Owner _____ Construction date _____

Date of first filling _____

Dates and duration of draw-downs _____

Kind of dam _____

Height of dam _____

Control structure _____

Effect of dam on fish movements _____

Water supply, source, adequacy, silt load, pollution _____

Bottom soil _____ Fertilization _____

Surrounding soil _____ Fertilization _____

Immediate shore _____

Surrounding land _____

Area _____ Max. depth _____ Ave. depth _____

Color _____ Turbidity _____

M.O. Alk. _____ pH _____

Vegetation _____

Fish food _____

Spawning facilities and success _____

Temperature _____ S _____ B _____

Thermal stratification _____

History of fish mortality —————————————————————————————
_____

Fish diseases and parasites _____

Fish species present _____

Stocking history _____

Fishing intensity _____

Annual fish harvest in kg by species _____

Recommendations (and other observations) _____

Observer _____ Date _____

Chapter 19

Shrimp Farming

Commercially Important Penaeid Species

Group	Origin	Common Name	Scientific Name
Cold water	North Atlantic North Pacific	Northern shrimp	*Pandalus borealis*
	NE Atlantic	Common shrimp	*Crangon crangon*
Tropical	Indo-Pacific	Green tiger prawn	*Penaeus semisulcatus*
		Banana prawn	*Penaeus merguiensis*
		Indian white prawn	*Penaeus indicus*
		Giant tiger prawn	*Penaeus monodon*
		Kuruma prawn	*Penaeus japonicus*
		Fleshy prawn	*Penaeus orientalis*
		Western king prawn	*Penaeus latisulcatus*
		Brown tiger prawn	*Penaeus esculentus*
	Western Indian Ocean	Indian white prawn	*Penaeus indicus*
		Giant tiger prawn	*Penaeus monodon*
		Green tiger prawn	*Penaeus semisulcatus*
	Eastern Atlantic	Southern pink shrimp	*Penaeus notialis*
	Western Atlantic	Northern white shrimp	*Penaeus setiferus*

Group	Origin	Common Name	Scientific Name
		Northern pink shrimp	*Penaeus notialis*
		Southern pink shrimp	*Penaeus notialis*
		Northern brown shrimp	*Penaeus aztecus*
		Southern brown shrimp	*Penaeus subtilis*
		Southern white shrimp	*Penaeus schmitti*
		Red spotted shrimp	*Penaeus brasiliensis*
	Eastern Pacific	Yellow leg shrimp	*Penaeus californiensis*
		White leg shrimp	*Penaeus vannamei*
		Blue shrimp	*Penaeus stylirostris*
		Crystal shrimp	*Penaeus brevirostris*
		Western white	*Penaeus occidentalis*
Freshwater	Indo-Pacific	Giant river prawn	*Macrobrachium rosenbergii*

Hatchery for the Giant Freshwater Prawn

The giant freshwater prawn, *Macrobrachium rosenbergii* is the largest of the freshwater prawn and is the most suitable species for culture. However, the seed resources are limited. This constraint can be mitigated through the establishment of backyard hatcheries.

M. rosenbergii hatcheries can be located at places with favorable climate conditions (25-30 degree Celsius temperature) where pollution-free sea water and freshwater are available with sufficient brood stock. Infrastructural facilities and technical support should also be available for establishment of a successful hatchery. The design of the hatchery depends on the site, production target and finantial inputs. Small scale hatcheries are more efficient and economical.

Brood stock maintained in freshwater ponds at a density of 2-3/sq.m. in the ratio of one male to four females attain maturity in 4-7 months, when fed with palletized feed containing 30 per cent protein. The species breeds throughout the year under favorable conditions and the average fecundity is 1000 eggs per gram body weight.

The berried females are released into larval rearing tanks with treated and matured sea water diluted to 11-13 ppt with clean freshwater. The larvae that hatch out are called zoeae and have eleven stages before metamorphosis. The larval rearing is the most critical stage and demands careful monitoring of the water quality parameters and continuous aeration.

A two phase larval rearing system with cylindro-conical tanks and flat-bottom tanks is the most economical one. Early larvae (zoea 1 to V) reared at 200-500 per liter in cylindro-conical tank are transferred to flat-bottom tanks at 1000/l for further development.

Feeding is started on the second or third day after hatching and for best results, palletized prepared feed with 40-50 per cent protein during day time and *Artemia* nauplii during night are recommended.

Rearing tanks are cleaned once a day and 20-25 per cent of water exchange twice a week. The water can also be recycled after exposing to sunlight for a week. Post-larvae normally start to appear between the 22nd and 32nd days and 90 per cent of the larvae metamorphose in the next 10 days. A week after sighting the first post-larva, the first cropping is done. Final harvesting is done a week after the first one.

Different Larval Stages of Giant Prawn

Stage	Age (Days)	Size (mm)	Character
I	1-2	2.0-2.1	Sessile eyes
II	3-5	2.2-2.4	Stalked eyes
III	6-8	2.5-2.7	Uropods appear
IV	8-13	2.8-3.1	Two dorsal teeth on the ostrum
V	12-19	3.2-3.4	Telson narrower and elongated
VI	15-24	3.5-3.6	Pleopod buds appear
VII	24-26	3.9-4.5	Pleopods biramous and bare
VIII	24-28	4.5-4.8	Pleopods with setae
IX	26-30	4.5-4.9	Endopods of pleopods with internae
X	29-33	4.8-5.0	3-4 dorsal teeth on rostrum
XI	32-35	4.9-5.2	Teeth on half of the upper dorsal margin
Post-larvae	34-51	5.3-6.0	Teeth on upper and lower margin of rostrum

Freshwater Prawn Culture

Preparation of Culture Ponds

Water with low acidity in the range of pH 6.8 to 7.5 is ideal for freshwater prawn culture. To lower the acidity, the water is usually treated with lime shell at the rate of 150 to 200 kg per acre. To certain extent this will eradicate most of the living organisms in the pond. Better results can be obtained by adding 1000 kg of mahua cake along with the treated lime shell. Within 15 to 20 days, the toxicity of mahua will fade away and the material will turn into useful fertilizer. Only then should the post-larvae of freshwater prawns be released into the pond.

If a continuous increase in acidity with a corresponding decrease in pH is noticed, several jute bags each containing 5 to 10 kg of powdered lime shell can be kept immersed by tying them to wooden poles fixed at various places. Occasionally it is good to shake the bags which will facilitate the release of melted lime into the water around and raise the pH. The problem of pH reduction can be fairly solved by this method.

Entry of brackish water which render the pond more salty will not cause any danger to the prawns as they can tolerate salinity up to 7 ppt. In fact, the increased salt content can prevent the growth of water lilies, weeds *etc.* which are likely to obstruct the growth of the prawns. These plants may be removed frequently from the ponds.

According to the topography of the place, inlets and outlets may be made for freshwater entry and exit of polluted water. If such provisions are not possible, water replacement can be done with pumps.

To the dried pond, prepared as explained earlier, one half to two tons of semi-dried cow dung mixed with10 to 15 kg of urea is broadcasted. 200 to 300 kg of poultry waste can also be used. Then water is raised and kept for 15 to 20 days prior to stocking of the post-larvae. Certain micro-organisms will grow in the manure to become a natural source of food for prawns.

Raising Juveniles from Post-Larvae (PL)

In the nursery, post-larvae is raised to juveniles and when they grow to a size weighing 3 to 4 grams, they are taken out to stock in culture ponds.

The nursery should have sluices with screen to facilitate easy to and fro flow of water. The PLs are released first to the nursery where they are allowed to grow to the juvenile size. A nursery of 100 cents area can hold up to 200000 PLs.

In the nursery when the bags containing post-larvae arrive, they must be kept in the water for 15 minutes to attain the temperature of the nursery water. The bags should then be opened and the water be allowed to mix slowly to avoid tension, that the PLs may likely to experience when they come in contact with the new environment. The PLs must be fed immediately after they are released since

Feeding Procedures in the Nursery

PL Size	Qty. of Food for 1000 PLs	Type of Food	Size of Food
1st-15th day	10 grams, morning and evening	Calm meat Grinded and cleaned	500 mesh
16th-30th day	15 g for 1st 5 days 20 g for next 5 days 25 g for remaining days	–do–	500-600 mesh
31st-45th day	50 g/day for 5 days 75 g/day for next 5 days 100 g/day for remaining days	a) Boiled shrimp meat, scrambled in the morning b) Evening-ground nut oil cake, 1000-2000 mesh size (with protein crackers, minerals, enzymes, vitamins fungal and viral antidotes every alternate day)	

they have been in the bags for a long time. In the nursery, the PLs are fed well. Treated water from the reservoir is permitted to flow into the nursery and an equal amount is moved out through the sluices. At least 10 per cent replacement of water, as described above, must occur daily. More replacement of water is always better.

Prawn Culture in Grow Out Ponds

Prior to the commencement of culture, the selected area must be dried and without disturbing the surface in between the trenches. Such trenches will provide shelter to the prawns. Cashew tree branches cut pieces, roofing corner tiles *etc.,* which all can be easily removed may be placed along the periphery of the pond to provide habitat to the prawns.

Once the juveniles are released into the large growing area, feeding should be in the evenings. Keeping the feed in large flat aluminum plates at several places around the pond is an easy way to fed. In large ponds low powered flat bottom fiber-glass boats are used to disperse the feed. This system is known as the broadcasting of feed.

For pond water management, the freshwater delivery canal is directed to a reservoir. In this reservoir all harmful organisms are destroyed and pH is regulated. This is achieved by immersing concealed clay pots filled with treated lime shells in several places around the pond. When the water enter the pots the lime shell will be reacted. The pressure due to the reaction will cause a sudden explosion of the pots killing most of the organisms which otherwise could pose a big threat to the prawns. The lime when gradually mixed with the water, the pH level will raise accordingly.

Stocking of juveniles in grow out ponds will vary from 3000 to 6000 per acre.

In the grow out ponds the most economical and productive feed should be a combination of ground nut oil cake, soyabin oil cake, fish meal *etc.* with protein crackers, minerals and enzymes, vitamins, fungal and viral antidotes. Frequent moulting will ensure that prawns are growing fast.

Every week-end a few prawns may be netted for random inspection to record their growth. If favorable conditions prevail in the pond, growth will be fast. Within 30 to 45 days, the young ones may attain a weight of over 15 g. The daily feed dispersed into the pond must amount to 10 per cent of the body weight of the growing prawns.

Prawn's Growth from the Time it is Released to Pond		The Feed Ration/1000 Prawns
From	*To*	
45th day	50th day	150 g
51st day	55th day	200 g
56th day	60th day	250 g
61st day	65th day	300 g
65th day	70th day	400 g
71st day	75th day	500 g

Prawn's Growth from the Time it is Released to Pond		*The Feed Ration/1000 Prawns*
From	*To*	
76th day	80th day	750 g
81st day	85th day	1000 g
86th day	90th day	1250 g
91st day	95th day	1500 g
96th day	100th day	1750 g
101st day	105th day	2000 g
106th day	110th day	2250 g
111th day	to harvesting date	2250 g

If more left over food is found, it can account for either an excess ration given or an increased rate of mortality.

If prawn culture is ventured following the above mentioned guidelines the prawn should grow up to 100-120 gram each within 5 to 6 months with a survival rate well over 85 per cent. To work out the economics, one may consider an average mortality rate of 40 per cent and an average growth of 50 gram each.

Chapter 20

Brackish Water Shrimp Farming

Selection of Healthy Fry for Farming

Shrimp fry that vary from light grey, brown to dark brown and black in color have consistently have high survival and growth rates. Post larvae that perform poorly are those that show signs of red or pink coloration which can be directly related to rearing or transport stress.

Good quality animals are found to be swimming actively. When in a basin, they may not always be moving, but will react to a tap on the side of the basin, movement in water or shadows.

Healthy fry feed actively and have a full digestive tract, unless there is no food available, such as, after a long shipment. Beware if fry at a hatchery is found with an empty gut. When animals are sick or stressed, they usually don't eat.

A clean shell indicates that the animal is growing fast and molting frequently. Slow growth is indicated by the presence of protozoans, other dirt and necrosis (black spot) on the shell.

Using a microscope, one can best see muscle development in the sixth tail segment. The muscle should completely fill the shell from gut down to the underside. In older fry (PL 22 and up), this may be pigmented and hard to see. When the PL is stressed, however, the muscle has a "grainy" look (much like the grain in wood) and is grayish or brown in color. Healthy fry have a clear smooth muscle.

Muscles are highly susceptible to stress, such as, rapid temperature and salinity changes. Even when fry appear healthy, alive and are feeding, the fry is stressed, if the tail muscle is grainy. If the muscle is clear and thick, then the fry also meet the other criteria; full gut, clean shell and good color.

Acclimation and stocking

Proper stocking techniques prevent unnecessary mortality. Easily measured pond water parameters, such as, temperature, salinity and pH are not the only factors to consider when stocking. Fry mortality caused by other variations in water quality can only be overcome by careful acclimation procedures. Acclimate the fry by adding pond water to a fry transport bag, or preferably in a large acclimation tank.

Temperature, salinity and pH are to be measured both of pond water and bag water. Empty bags containing fry into basin or acclimation tank. Pond water at the rate of 250 ml per minute in case of one bag containing 10 liters of water be added.

Management Practices for Shrimp Farming

Stock Manipulation and Management

Sound stocking practices is a balance between the stock and available food. Stocking rate directly depends upon rate of supplementary feeding, although to some extent depends upon available natural food in extensive culture system.

Production and Maintenance of Natural Food

Production of natural food can not be ignored completely especially in nursing phase. Natural food is grown by using a combination of organic and inorganic fertilizers. Kind of natural food grown depends upon species cultured and even differs to different growth stages. Lab-lab is good for nursing prawns, but not for grow-out prawns. Plankton is natural food base grown in all systems of prawn culture.

Plankton Production

These procedures are required to be followed.

1. Drain and completely dry the pond bottom
2. Apply lime at the rate of one ton per hectare.
3. Admit water to a depth of 30 cm.
4. Apply organic fertilizers (cow dung or chicken manure) placed in plastic sacs tied and submerged in water like a tea bag (T-bag). About 16-20 sacks containing 3 kerosene cans per bag are needed per hectare for cow dung manure and 10 sacks containing 2 kerosene cans per bag are needed per hectare using chicken manure.
5. Fertilize with 20 kg urea and 20 kg diammonium phosphate per hectare. Planktons such as, diatoms or brown algae should grow after 3-5 days. Absence of phytoplankton growth after 5 days indicates the need for more fertilizer. Add one-half the amount recommended every 2 days thrice or four times until the pond water turns brown. When the brown color is attained, fill the pond with water to a depth of 50 cm. Gradually increase the depth of water to 80 cm. Test the transparency of water with a Secchi disc. If transparency registered a minimum of 20 cm or a maximum of 34

cm, the pond is ready for stocking. Remove T-bags, place them temporarily on top of the dyke and reuse them once transparency increases.

Maintenance of Natural Food Growth

Stocking over a period of time removes higher amount of nutrients from the soil, thus depleting the soil of pond nutrients needed for the growth of natural food. If these nutrients are not replaced, the crop will be affected, thus the need for fertilization.

For plankton, close monitoring to water transparency is required. If the transparency registers less than 20 cm. immediately drain and replenish about 30 per cent of the water volume to prevent the danger of plankton bloom which cause oxygen depletion and mass mortalities. If transparency registers over 35 cm. plankton growth is depleted, hence, reuse T-bags and apply 5 kg of urea and 5 kg of diammonium phosphate. Remove T-bags once transparency returns to recommended levels.

Feeds and Feeding

The application of feed is a vital operation that should be done properly, if not pollution of culture water and wastage of feed will result.

Water Quality Management

Water quality is one of the most important factors that affect prawn production. Good environment is necessary for optimum growth and survival.

Dissolved Oxygen

The desired level of dissolved oxygen of 3 to 10 ppm in the pond indicates that the pond system is functioning efficiently. Sudden change of weather condition causes oxygen depletion. Plankton bloom greatly reduce photosynthetic input of dissolved oxygen in the pond.

The desired color of pond water is light brown. The desired transparency depth is of 20 cm minimum and 30 cm maximum. Water should be changed immediately if the transparency is less than 20 cm.

Water Temperature

The temperature of water affects the metabolism of prawn and indirectly, the growth rate. The ideal water temperature for shrimp farming at different stocking densities are;

Water depth – 60 to 90 cm Shrimp seed stocking – 5000 to 90000

Water depth – 100 to 150 cm, Shrimp seed stocking – 100000 to 500000

DOC	15	30	45	75	90	105	120	135	150
					(Temp. °C)				
5000	32	31	31	30.5	30.5	30	30	30	30.5
10000-30000	32	31	31	30.5	30.5	30	30	30	30
40000-50000	31.5	31	30.5	30.5	30.5	30	30	30	30
60000-70000	30.5	30.5	30.5	30	30	30	30	30	30
80000-90000	30.5	30.5	30	30	30	30	30	30	30
100000-150000	30	30	30	30	30	29.5	29.5	29.5	29.5
200000-250000	30	30	29.5	29.5	29	29	29	29	29
300000-350000	30	29.5	29	29	29	28.5	28.5	28	28
400000-450000	30	29	29	28.5	28.5	28	28	27.5	27.5
500000	30	29	28.5	28.5	28	28	27.5	27.5	27.5

DOC: Days of Stocking.

pH

The pH of pond water should be within the range of 6.5 to 8.5. Above or below this range the water should be changed.

Hydrogen Sulfide

At a high concentration of 2 ppm, the shrimps will die. Hydrogen sulfide concentration could be reduced by changing the water and by drying the pond bottom.

Detection of Disease Out-breaks in Shrimp Ponds

To detect the onset of diseases early, careful monitoring of prawns is essential. The appearance and behavior of shrimps may give clear signs of onset of diseases. For early detection of diseases in shrimp ponds, the following signs of symptoms can serve as guidelines.

1. Loss of appetite
2. Abnormal changes in color
3. Exoskeletal rot/lesion/erosion
4. Physical deformity
5. Opaque musculature
6. Abnormal swimming
7. Abnormal condition/color of gills
8. Lethargy
9. Slow growth
10. Increased exoskeletal epiboint

11. Abnormally prolonged soft-shelling

12. Mortalities

In case of disease outbreaks without any characteristic manifestation the shrimp farmer can narrow down on the probable cause of the disease by observing the mortality pattern.

a) Gradual increasing mortalities over several days to weeks are usually caused by micro-organisms (virus, bacteria, fungi parasites) or nutritional deficiencies.

b) Sudden mass mortalities are generally associated with adverse physico-chemical parameters (low dissolved oxygen, acidic soil, high ammonia/nitrite levels, thermal shock, toxic substances *etc.*)

For detecting the causes, regular monitoring and recording of physico-chemical parameters like temperature, turbidity, dissolved oxygen, salinity, ammonia, pH and nitrite and checking pond bottom conditions, feed consumption etc are very useful.

Pen Culture of Tiger Shrimp – *Penaeus monodon*

For selecting suitable site for pen culture of tiger prawns, following points are required to be considered.

1. Average water salinity should be within the range of 10 to 32 ppt.

2. Bottom sediment should be clayee.

3. The slope of the bottom should not exceed more than 15 degrees.

4. The area should be protected from strong winds, waves and currents.

5. Water depth should be about 2 meter at low tide.

6. The tidal fluctuations should allow the water depth to be at least one meter at low water spring tide.

7. The site should be relatively free from domestic, industrial and agricultural wastes and other environmental hazards.

8. The site should have enough water circulation to improve on poor water quality that could occur at some stage in culture.

Eradication of Pest and Predators

Tea seed cake is a by-product from the processing of the plant *Camelia* sp. The seeds of the plant contains saponin, a chemical, which is the effective ingredient for poisoning pest and predators in fish ponds. It has no effect on invertebrates.

During Pond Preparation

1. After sun drying, plowing and liming of pond, fill the pond to 10-20 cm water.

2. Tea seed cake should be properly pulverized to effect good water solubility.

3. Soak or dissolve tea seed cake in larger container with water.

4. Apply solution and mix thoroughly to the pond water.

5. Let stay overnight.
6. Fill up pond to suitable stocking depth the following day.
7. Calculate the amount of tea seed cake at 15 to 20 ppm.

Fish Eradication during Culture Period

1. A single treatment of tea seed cake during a crop cycle should be sufficient for cost efficiency. This should be scheduled between the 60th to 75th day of the crop or when shrimps have attained at least 10 g average body weight unless when absolutely necessary.
2. Drain at least 50-60 per cent of pond water.
3. Spread hydrated lime on the dike at the water's edge all around the pond as soon as desired depth has been attained.
4. Soak or dissolve tea seed cake in large container with water.
5. Apply and mix thoroughly to pond water. Application should be made during the most suitable period of the day (around 10 A.M.). The preceding procedures should be properly timed or scheduled wherein the events are immediately successive.
6. Paddle wheel should be operated initially for two hours during the exposure period.
7. Abort treatment if rain occurs or cloudy skies predominate.
8. Treatment should take effect within 4-6 hours after application of tea seed cake. Water should not be added during this period.
9. Refill pond to normal water depth.
10. Do not expose shrimps overnight.
11. Calculate amount of tea seed cake at 15 to 25 ppm. If cloudy, use the higher concentration. If the salinity is low, use the higher concentration.

Aeration is a must for maintaining the dissolve oxygen level, especially in semi-intensive shrimp culture.

Aeration schedule of ponds stocked with 50000/ha and 150000/ha from the day of stocking until harvest are given below.

Days of Culture	Aeration Time	Total Aeration Time/Day
1–30	Emergency use only	
31–45	11 p.m.–7 a.m.; 12 noon–3 p.m.	11 hours
46–60	9 p.m.–7 a.m.; 12 noon–3 p.m.	13 hours
61–75	7 p.m.–7 a. m. ; 12 noon–3 p.m. 15 hours	
76–90	6 p.m.–7 a.m.; 12 noon–3 p.m. 16 hours	
91–105	5 p.m.–7 a.m.; 12 noon–3 p.m.	17 hours
106 – 120	12 noon–10 a.m.	22 hours
121 – till harvest		24 hours

Production and Harvest

The estimated gain in average body weight, total production, feed conversion ratio and survival at harvest in two stocking densities (50000/ha and 150000/ha) are;

Stocking Density	Av. Body Weight at Harvest (g)	Total Production (kg/ha)	Survival (Per cent)	FCR
(50000/ha)	31.23	1278.80	81.90	1.47
"	28.09	1352.00	96.25	1.20
"	26.47	1104.00	93.40	1.30
Mean	28.60	1244.93	93.40	1.35
(150000/ha)	28.30	3690.10	86.90	1.90
"	27.79	3546.01	93.40	1.92
"	28.78	3444.00	80.78	1.57
Mean	28.29	3560.04	87.03	1.79

Shrimp Feeding Management

The objectives of feeding management are, (a) to prevent overfeeding of the stock which can be the cause of serious water problem and (b) to avoid underfeeding as this can reduce income due to slow growth of shrimp and longer culture period.

Since the feed represents 70-80 per cent of the total shrimp production cost, a nutritious and economical diet is always a worthy investment.

At different stages, shrimps are fed with specific amount and type of feed to satisfy their nutritional requirement.

Shrimps are nibblers and slow feeders. They take the feed with their pinchers and bring this to their mouth and slowly chew the feed. If the feed is small enough, they will throw it into the mouth.

Major Nutrients Needed by the Shrimp

35-45 per cent protein, 25 per cent carbohydrate, 10 per cent fat, besides vitamins and minerals. The vitamins are necessary for proper utilization of protein, carbohydrate and fat.The minerals like calcium and phosphorus are essential for formation of exoskeleton or cell. All these nutrients are so inter-related that they have to be incorporated in the diet to be fully utilized by the body.

Feed manufacturers have different types of names for shrimp pellets given at different stages of growth. But generally the pellets are classified in three categories, namely, starter (0.25-0.5 mm), grower (1.5-1.75 mm) and finisher (1.75-2.2 mm).

Prawn feeds are usually packed with plastic lined container to minimize the uptake of moisture and to help protect the flavor, color and aroma (attractant).

It is very important that the shrimp grower knows the optimum feeding time and frequency, feeding rate and methods of determining rate for shrimps. Feed manufacturers have their own feeding times and rates.

Feeding rate is the amount of feed to be fed daily and is based on percentage of body weight of the prawn. The computation of the feeding rate is also based on survival and actual feed consumption.

Feeds are broadcasted evenly throughout the pond and a certain amount is placed on the feeding trays. The actual number of feeding trays depends on the actual experience of the grower. The usual number ranges from 8 to 10 trays per pond regardless of the area. The consumption of actual feed is observed within one to two hours after feeding.

It is suggested that feeding should commence one or two days after stocking. For a reason that in an intensive culture system, the pond water provides very little amount of natural food to the cultured shrimp and early feeding will provide early training for the shrimp to take pellets. With this method, higher survival is assured.

The frequency of feeding varies with the size of shrimp and of the days of culture. Shrimp growers can use this as reference for daily feeding from the day feeding commenced until harvest.

Stock Sampling Method and Number to be Sampled

Shrimps are sampled using different methods. Samples can be taken from feeding trays (from day 1 to 45-60 days) and with cast net (day 61 to harvest). Data gathered from this sampling will be used in adjusting daily feed requirement, survival and to determine the incidence of diseases or parasites. The amount or number of samples varies with the age of shrimp or days of culture.

Target Growth, Survival and Feeding Rates of Shrimp in Semi-Intensive System

Culture Period	Estimated Growth (g)	Survival (Per cent)	Per cent Body Wt. of Feed Daily
7	0.03	100	30
15	0.21	100	20
30	1.70	95	15
45	5.90	95	10
60	10.50	90	8
67	13.40	85	8
72	15.00	85	6
80	18.20	80	6
86	21.00	80	5

Culture Period	Estimated Growth (g)	Survival (Per cent)	Per cent Body Wt. of Feed Daily
90	23.00	75	5
97	27.00	75	4
105	31.00	75	4
112-120	36-40	75	3

Summary of Growth, Feeding Rate and Frequency and Amount of Feed to be given Daily for a Stocking Density of 200000 nos/ha.

DOC	Expected ABW (g)	Feed to be given per Day (g)	Type of Feed per Day	Feeding Frequency
0-15	0.24	5.0	Starter	3x
16-30	1.70	13.0	Starter	3x
31-45	5.90	25.0	Grower	4x
46-60	10.50	45.0	Grower	4x
61-75	15.00	75.0	Grower	4x
76-90	21.00	120.0	Finisher	5x
91-105	26.00	180.0	Finisher	5x
106-112	31.00	200.0	Finisher	5x
112-121	34.00	214.0	Finisher	5x

Drug for Shrimp Diseases

Prefuran (Furanace, Nifurpirinol, P-7138) is the only aquatic drug absorbed from bathing water providing relief from most bacterial infections, fungal infections and parasitic protozoal infections. The drug can be used on all edible shrimps both fresh and marine and most of the infections have been cured with single treatment unlike traditional antibiotics, which is required to be administered for 7 to 10 days. The drug is rapidly absorbed, metabolized and excreted from shrimps and leaves no detectable tissue residues after bathing in freshwater for several hours. For administration of the drug the following rules are required to be observed.

1. Temperature of treatment bath should be similar to habitat.
2. Aerate treatment solutions to maintain a favorable dissolved oxygen level.
3. Test several shrimps before doing large groups.
4. Observe the dosages outlined below.
5. If edible shrimp are to be treated, a minimum 5 day withdrawal before consumption is suggested.
6. The dose of the drug is recommended as;

Treatment – I

3 to 5 days bath at 0.05 to 0.1 ppm of active ingredient. The following table may be used to calculate grams of PREFURAN per liter of water.

Water Volume (Liters)	Grams of Prefuran
100	0.05 to 0.10
200	0.10 to 0.20
400	0.20 to 0.40
600	0.30 to 0.60
800	0.40 to 0.80
1000	0.50 to 1.0

Treatment – II

15 to 30 minutes bath at 1.0 to 2.0 ppm of active ingredient.

Water Volume (Liters)	Grams of Prefuran
100	1.0 to 2.0
200	2.0 to 4.0
400	4.0 to 8.0
600	6.0 to 12.0
800	8.0 to 16.0
1000	10.0 to 20.0

It is effective for the disease mentioned below.

1. Bacterial infection of Aeromonas
2. Bacterial infection of Vibrio
3. Bacterial infections of Flexobactor columnaris
4. Fungal infections of Saprolegnia
5. Parasitic infections of Ichthyopthirius
6. Multifillis "Ich" or white spot
7. For general prophylaxis of surface abrasions, wounds and lesions.

References

Alikunhi, K. H. Standardization of biological studies in fish culture research; Proc. of the FAO World Symposium on Warm-water Pond Fish Culture, *FAO Fisheries Report* No. 44, Vol. V, 1968.

Butler, R. L. Fish behavior observation techniques; Workshop Proc. Biological Services Program, FWS/OBS-78/30, Fish and Wildlife Service, U.S. Department of the Interior, 1978.

Biswas, K. P. Fisheries Manual, M. Biswas, Kolkata-84, 1993.

Chaudhuri, H. Breeding and selection of cultivated warm-water fishes in Asia and Far East, Proc. of the FAO World Symposium on Warm-water Pond Fish Culture, *FAO Fisheries Report* No. 44, Vol. V, 1968.

Hocutt, C. H. Fish, Workshop Proc. Biological Services Program, FWS/OBS – Fish and Wildlife Service, U. S. Department of the Interior, 1978.

Isom, B. G. Benthic Macro-invertebrates, Workshop Proc. Biological Services Program, FWS/OBS- Fish and Wildlife Service, U. S. Dept. of the Interior, 1978.

Lagler, K. F. Studies in Freshwater Fishery Biology, J. W. Edward, Ann Arbor, Michigan, 1949.

Mason, T. W. Jr. Methods for the Assessment and Prediction of Mineral Mining Impacts on Aquatic Communities, Workshop Proc. Biological Services Program, FWS/OBS- Fish and Wildlife Service, U. S. Department of Interior, 1978.

Prowse, G. A. Standardization of Statistical Methods in Fish Culture Research, Proc. of the FAO World Symposium on Warm-water Pond Fish Culture, *FAO Fisheries Report* No. 44, Vol. V, 1968.

Polgar, T. T. Statistical Approach to Biological Predictions, Workshop Proceedings, Biological Services Program, FWS/OBS- 78/30, Fish and Wildlife Service, U.S. Department of the Interior, 1978.

Preston, H. R. Bioassay Techniques, Workshop Proceedings, Biological Services Program, FWS/OBS – 78/30, Fish and Wildlife Service, U. S. Department of the Interior, 1978.

Raleigh, R, F. Habitat Evaluation Procedure for Aquatic Assessments, Workshop Proceedings, Biological Services Program, FWS/OBS – 78/30, Fish and Wildlife Service, U. S. Department of the Interior, 1978.

Raschke, R. L. Macrophyton, Workshop Proceedings, Biological Services Program FWS/OBS – 78/30, Fish and Wildlife Service, U. S. Department of the Interior, 1978.

Swingle, H. S. Standardization of chemical analysis for waters and pond mud, Proc. of the FAO World Symposium on Warm-water Pond Fish Culture, *FAO Fisheries Report* No. 44, Vol. V, 1968.

Spoon, D. M. Aufwuchs, Workshop Proceedings, Biological Services Program, FWS/OBS – 78/30, Fish and Wildlife Service, U. S. Department of the Interior, 1978.

Sage, L. E. Zooplankton, Workshop Proceedings, Biological Services Program, FWS/OBS – 78/30, Fish and Wildlife Service, U. S. Department of the Interior, 1978.

Weber, C. I. Phytoplankton, Workshop Proceedings, Biological Services Program, FWS/OBS – 78/30, Fish and Wildlife Service, U. S. Department of the Interior, 1978.

Index

www.ingramcontent.com/pod-product-compliance
Lightning Source LLC
Chambersburg PA
CBHW031951180326
41458CB00006B/1691